Krouse, John K.,
1947-

What every engineer should know about computer-aided design and computer-aided manufacturing

DATE			

What Every Engineer Should Know About

Computer-Aided Design and Computer-Aided Manufacturing

WHAT EVERY ENGINEER SHOULD KNOW

A Series

Editor

William H. Middendorf

Department of Electrical and Computer Engineering
University of Cincinnati
Cincinnati, Ohio

Vol. 1 What Every Engineer Should Know About Patents, *William G. Konold, Bruce Tittel, Donald F. Frei, and David S. Stallard*

Vol. 2 What Every Engineer Should Know About Product Liability, *James F. Thorpe and William H. Middendorf*

Vol. 3 What Every Engineer Should Know About Microcomputers: Hardware/ Software Design: A Step-by-Step Example, *William S. Bennett and Carl F. Evert, Jr.*

Vol. 4 What Every Engineer Should Know About Economic Decision Analysis, *Dean S. Shupe*

Vol. 5 What Every Engineer Should Know About Human Resources Management, *Desmond D. Martin and Richard L. Shell*

Vol. 6 What Every Engineer Should Know About Manufacturing Cost Estimating, *Eric M. Malstrom*

Vol. 7 What Every Engineer Should Know About Inventing, *William H. Middendorf*

Vol. 8 What Every Engineer Should Know About Technology Transfer and Innovation, *Louis N. Mogavero and Robert S. Shane*

Vol. 9 What Every Engineer Should Know About Project Management, *Arnold M. Ruskin and W. Eugene Estes*

Vol. 10 What Every Engineer Should Know About Computer-Aided Design and Computer-Aided Manufacturing, *John K. Krouse*

Other volumes in preparation

What Every Engineer Should Know About

Computer-Aided Design and Computer-Aided Manufacturing

The CAD/CAM Revolution

John K. Krouse
Machine Design Magazine
Cleveland, Ohio

MARCEL DEKKER, INC. New York and Basel

Library of Congress Cataloging in Publication Data

Krouse, John K., [date]
What every engineer should know about computer-aided design and computer-aided manufacturing.

(What every engineer should know ; v. 10)
Bibliography: p.
1. Engineering design—Data processing.
2. Production engineering—Data processing.
I. Title. II. Series.
TA174.K76 1982 620'.0028'54 82-9957
ISBN 0-8247-1666-3 AACR2

MARCEL DEKKER, INC.
270 Madison Avenue, New York, New York 10016

Current printing (last digit):
10 9 8 7 6 5 4 3 2 1

PRINTED IN THE UNITED STATES OF AMERICA

For their love, support, and advice,
this book is dedicated to my family:
Louisa, Susanne, and James.
They are with me always
and are my highest inspiration.

Contents

Preface

Computer-aided design (CAD) and computer-aided manufacturing (CAM) are affecting almost every area of engineering. The number of CAD/CAM installations is growing by more than 40% a year. By 1985, 90% of all mechanical drafting is expected to be done by CAD, and about 30% of all manufacturers probably will use some form of CAM.

Most experts agree that computer systems such as these will continue to do what they have always done: free engineers from the tedious, time-consuming tasks that have nothing to do with technical ingenuity. Experience shows that CAD/CAM speeds the engineering process, stripping away the drudgery and paper-work that inhibits productivity and creativity.

The National Science Foundation Center for Productivity states that "CAD/CAM has more potential to radically increase productivity than any development since electricity." And CAD/CAM is viewed by many as the key to improving manufacturing productivity and the best approach for overcoming the worldwide economic slump.

Even though CAD/CAM is a relatively new technology, it already is having a huge impact on engineering. Moreover, the capabilities and impact of future systems are sure to be even greater. Changes in engineering usually are slow and evolutionary. Even radical innovations usually cause no more than a ripple in engineering as a whole. But CAD/CAM is exploding in every corner of the profession like nothing before, and its effects are being felt in every phase of engineering. If there was ever anything in engineering that could be called a revolution, this is it.

CAD/CAM will magnify man's mental power just as the machines of the industrial revolution expanded the strength of his muscles. This wave of change, however, will depend not on exhaustible natural resources but on the limitless creativity of the human mind. And we are now just seeing the beginning of this new approach to engineering.

Engineers and managers trying to learn about this new technology often encounter obstacles. There is an abundance of literature on specific CAD/CAM systems and fragmented areas of computer technology. But not many well-rounded experts know how all the diverse areas fit together. And few can explain CAD/CAM in a vocabulary that novices can understand easily. In addition, there are fundamental problems in terminology and definition because CAD/CAM means different things to many people. Some regard CAD/CAM as automated drafting and NC tape preparation. Others include virtually all tasks performed with a computer as CAD/CAM. Engineers and managers clearly need explanations to put the picture into better focus.

The need for clarification is the inspiration for this book, which is intended to be a helpful guide to important activity in CAD/CAM. The book pinpoints and defines major CAD/CAM areas, shows how they fit together, describes major CAD/CAM users, and outlines the cooperative efforts to develop unified systems. The book is meant to serve as a solid base on which the novice may build further knowledge about CAD/CAM. And it may provide a revealing perspective of CAD/CAM even for those intimately involved with the technology.

John K. Krouse

About the Author

John K. Krouse is Staff Editor of *Machine Design Magazine.* He received the B.S. degree (1969) in physics from Case Institute of Technology, Cleveland, Ohio. Mr. Krouse's experience as an engineer involved designing control systems for guided missiles, torpedoes, and nuclear power systems at Vitro Laboratories, Gould Inc., and Bailey Controls Co. (1969–76). He currently specializes in CAD/CAM, and has published numerous articles on the subject. Mr. Krouse has won awards for editorial excellence from the American Society of Business Press Editors for his articles *Stress Analysis on a Budget* and *CAD/CAM: Bridging the Gap from Design to Production.*

1
Birth and Growth of CAD/CAM

For a long time, computers were bulky, expensive machines that could be operated only by those familiar with programming and related tasks. But dramatic technical improvements have made them smaller, more powerful, less expensive, and much easier to use. As a result, computers have proliferated into many diverse areas and have transformed our world.

Business, education, medicine, science, and other sectors of our society have benefited from the computer. But the single area depending most on the computer and experiencing the most dramatic growth is CAD/CAM—the combined technology of computer-aided design (CAD) and computer-aided manufacturing (CAM). This skyrocketing technology is the result of years of development in which computer systems have continually evolved. Early work nearly forty years ago laid the foundation for today's CAD/CAM systems.

Digital computers first appeared in the 1940s. They were huge, electromechanical machines that used clicking relays to perform computations. The first and largest of this generation of computers was the 5-ton Automatic Sequence Controlled Calculator, or the MARK 1, shown in Figure 1.1. Electromechanical computers performed calculations much faster than earlier mechanical calculators and computers. The MARK 1, for example, could add or subtract two 23-digit numbers in 0.3 second. And it could multiply them in 6 seconds.

Soon relays and other mechanical moving parts were replaced by vacuum tubes. Electronic *flip-flop* circuits made of vacuum tubes were switched on and off like switches with electronic pulse signals. The first of these vacuum-tube computers was ENIAC (Electronic Numerical In-

Figure 1.1. The first and largest electromechanical computer ever built was the MARK 1. Completed in 1944, the computer weighed 5 tons and contained over 3,300 relays connected by 500 miles of wiring. (Courtesy of IBM Corp., White Plains, New York.)

tegrator and Calculator) developed for the U.S. Army in 1946. Subsequently, many other vacuum-tube computers emerged in rapid succession in the late 1940s. One of the most powerful of these was IBM's Naval Ordnance Research Calculator shown in Figure 1.2.

Because the electronic pulses and circuits reacted thousands of times faster than mechanical parts, computation speed increased dramatically. In the mid-1940s, a vacuum-tube computer multiplied two 10-digit numbers in 1/40 second. By the mid-1950s, this was done in 1/2,000 second. Computing speed also increased with the advent of magnetic devices such as disks and drums for data storage, and magnetic cores for program memory. These replaced the punched cards and other manual data-entry methods of earlier computers.

Transistors replaced vacuum tubes in computer circuits in the late 1950s, creating a new generation of faster, more compact equipment. The most powerful computer of the day was IBM's STRETCH computer shown in Figure 1.3. The use of these solid-state devices changed computers dramatically. The transistor was 1/200 the size of a vacuum tube and could be packaged tightly because it produced only a fraction of the heat of vacuum tubes. Thus, computers became much more compact. Furthermore, signals had less distance to travel and response time of the solid-state devices was rapid; so computing speed also increased. Transistorized computers could multiply two 10-digit numbers in 1/100,000 second. In addition, solid-state transistors were far more rugged and reliable than vacuum tubes.

Computer size and cost were further reduced and computing speeds increased in the 1970s with integrated circuitry. Entire circuits consisting of thousands of transistors and other components were condensed onto confetti-size silicon chips. One of these chips from a computer-logic circuit is shown in Figure 1.4. The speed at which these chips switch signals permit the computer to make millions of calculations per second. It also reduces the size of the computer to a fraction of that normally required when circuits are made of discrete components.

Because of integrated circuit technology, computers today are much

Figure 1.2. The most powerful computer in the mid-1950s was the Naval Ordnance Research Calculator. It contained 9,000 vacuum tubes. (Courtesy of IBM Corp., White Plains, New York.)

Figure 1.3. Built in 1960, the powerful STRETCH computer contained 150,000 transistors and could execute 100 billion instructions per day. (Courtesy of IBM Corp., White Plains, New York.)

more compact, faster, and less expensive than their earlier counterparts. The largest mainframe computers now manipulate huge amounts of data, and their power dwarfs that of earlier equipment. Furthermore, smaller minicomputers about the size of a desk now handle complex tasks such as stress analysis that only a few years ago required a mainframe. And typewriter-size desktop computers like the one in Figure 1.5 contain the computing power of a bulky system that would have filled a room years ago. The smallest is the so-called microcomputer that has all the necessary circuitry for elementary computations on a clipboard-sized printed-circuit board containing a few integrated circuit chips and other components. This dramatic compression of computer circuitry has given built-in intelligence to products such as ovens, tools, cash registers, and automobiles.

This entire range of computers—from giant mainframes to tiny microcomputers—are a part of CAD/CAM. Much of the equipment in a CAD/CAM system, such as intelligent terminals and plotters, have built-in microcomputers that assume some of the computing burden in the overall system and give peripheral equipment greater independent capabilities. Desktops are now being used with larger host computers for performing computations in sophisticated networks. Most turnkey systems that make up the greatest portion of CAD/CAM systems use minicomputers. And powerful mainframes perform structural analysis, manipulate huge matrices of data, and complete other complex tasks in the most sophisticated computer systems.

Researchers are still miniaturizing circuitry even further for faster, more compact computers of the future. Some experts even envision that analysis by the year 2000 may be performed on hand-held computing modules shown in Figure 1.6. These units are predicted to be so compact and economical that they will be used routinely by engineers much the same as pocket calculators are used today.

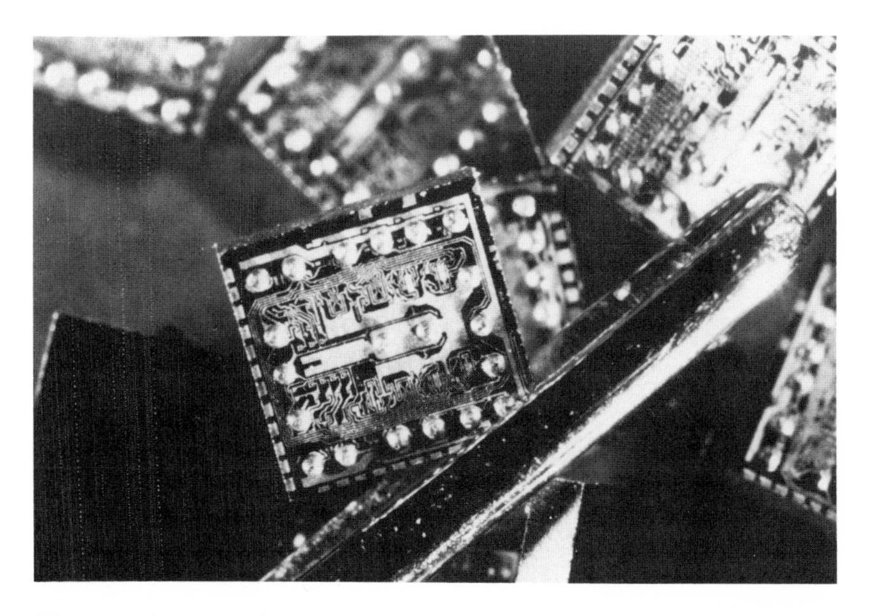

Figure 1.4. Contained in the eye of a needle, this silicon chip has miniaturized circuitry that switches a computer's electronic signals in a matter of nanoseconds. (Courtesy of IBM Corp., White Plains, New York.)

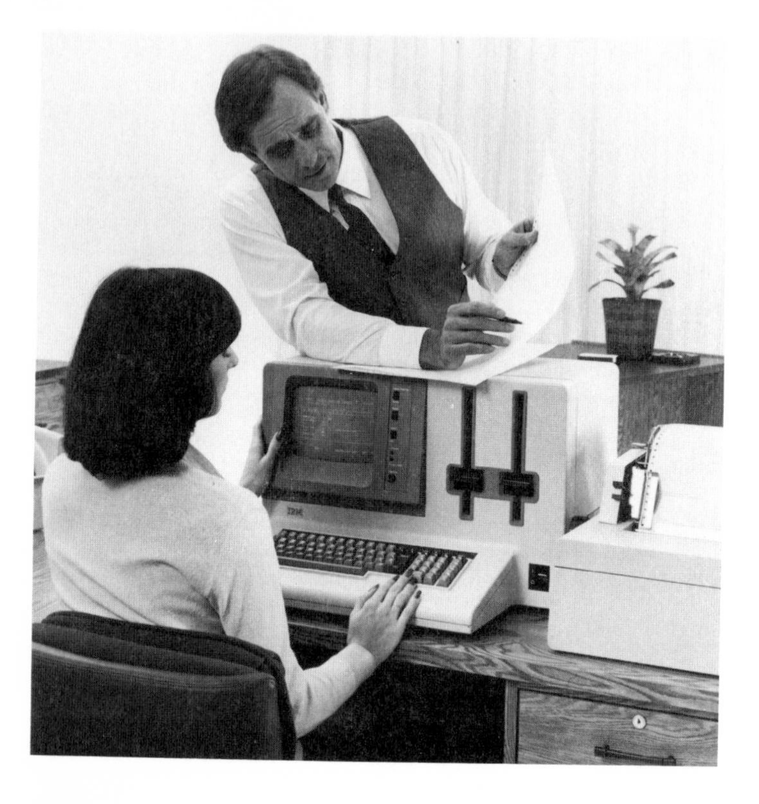

Figure 1.5. The IBM Model 5120 desktop computer has the power of equipment that 20 years ago filled a 20-by-30-foot room and weighed a ton. (Courtesy of IBM Corp., White Plains, New York.)

INTERACTIVE GRAPHICS

Computers are more compact, less expensive, and more powerful than ever. But one of the major reasons for the skyrocketing proliferation of CAD/CAM systems is the increasing ease with which a user communicates with the computer. Formerly, the user entered data and instructions into the computer with stacks of coded punched cards or reels of tape. And results were extracted from the computer in columns of raw numbers printed out on reams of perforated paper. Thus, the user had to be experienced in computer programming and operation to use the machine. And considerable time was usually required to execute a program and interpret the results.

But these early limitations of computer operation (which still prevail today in many people's minds when they think of a computer) were eliminated with the advent of interactive graphics. Here, the user communicates with the computer in display-screen pictures. Virtually no knowledge of computers is required to operate these so-called friendly systems. Furthermore, the communication between man and computer is carried out in real time. That is, the computer's response to its user's instructions is almost instantaneous.

An example of interactive computer graphics is shown in Figure 1.7. The terminal displays the results of a stress gradient analysis as lines of constant stress. Areas of high stress are easily pinpointed as those with high concentrations of the lines. The long list of numerical data which this

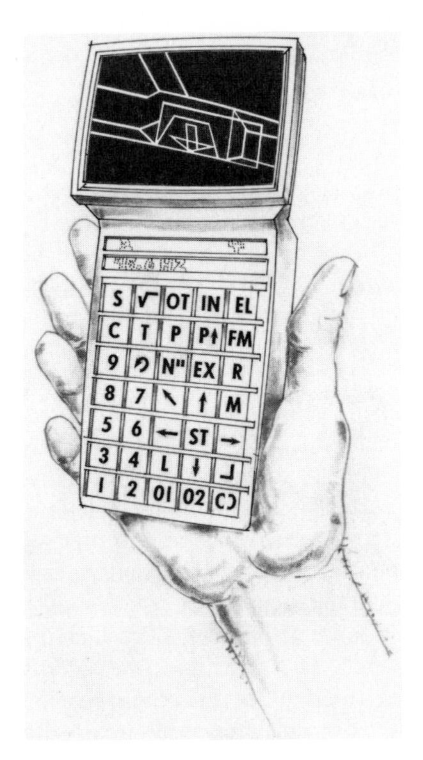

Figure 1.6. Future computer systems may be accessed through hand-held modules as compact and economical as today's pocket calculators. (Courtesy of Structural Dynamics Research Corp., Milford, Ohio.)

Figure 1.7. Concentration of stress lines easily pinpoints areas of potential fatigue in this plot displayed on a Tektronix Model 4054 interactive graphics terminal. The vast amount of data behind the plot would be tedious to interpret in tabular form. (Courtesy of Tektronix Inc., Beaverton, Oregon.)

plot represents would have been virtually indecipherable in tabular form.

Interactive graphics evolved as a means of conveniently translating man's mental images into computer language and vice versa. Internally, the computer handles its data in binary form. That is, all numbers, letters, symbols, and instructions are represented as a sequence of two digits—zeros and ones. A single letter of the alphabet, for example, might be represented by 11010001. This is called machine language.

Machine language is in binary form because of the ease with which the data can be manipulated in the computer. The zeros and ones of the machine language are simply represented by circuits in the computer being either off or on. In today's computers, compact integrated circuits are switched off and on. But the same basic principle was used to represent data in earlier computers with individual transistors, vacuum tubes, or electromechanical relays.

In the early days, a user communicated with the computer directly in machine language. That is, he wrote out long lists of zero-and-one sequences to execute even a simple command such as multiplication or division.

However, by the mid-1950s programming was simplified with symbolic coding. The programmer used a set of English-language statements that stood for certain standard computer commands. These statements were then converted automatically into machine language and fed into the computer by a translation program. One of the first high-level programming languages was FORTRAN (Formula Translation) developed for scientific and engineering data and still in wide use today. Other languages such as COBOL and BASIC for business applications also appeared and are still used.

These programming languages increased the speed with which a user could interact with the computer. But programming training skill at operating the computer and familiarity with the internal workings of the machine were generally required. And the lengthy numerical printouts containing the results from the computer were tedious and time-consuming to interpret.

Then in the early 1960s, interactive graphics was developed to permit a user to reap the benefits of the computer without training in programming, time-consuming coding, or other computer tasks. One of the earliest developments in interactive graphics was the Sketchpad Project at the Massachusetts Institute of Technology. In this system, a cathode-ray tube (CRT) display scope, similar to those used at that time to identify aircraft, was connected to a computer. Data was entered with a hand-held lightpen, which was a cylinder with a photocell attached to the end. The computer sensed the position of the lightpen on the scope and responded by lighting that point on the scope and storing the coordinate data in its memory.

By specifying points on the scope and executing simple computer commands, the user could quickly generate straight lines, circles, arcs, and other geometries. Producing a circle, for example, required the user only to specify the radius and the location of the center. With this technique, the user could quickly produce an entire diagram on the display screen. And the data base of coordinates stored in the computer could subsequently be used to manipulate the display image, produce hard-copy drawings, or be entered as an input to some form of geometric analysis. A feature that made interactive graphics so appealing was that the communication with the computer was carried on in real time. Thus, there was no long waiting time for the computer to grind through batch calculations. Feedback from the computer was almost instantaneous, permitting the interaction to take place almost in a conversational mode.

Several interactive graphics systems were developed in the 1960s. However, their use was restricted mostly to very large companies such as General Motors and Boeing that developed their own in-house systems. These companies could afford the development costs of the system and the expensive mainframe computers required to operate the systems. By the early 1970s, however, interactive graphics could be performed on less-expensive minicomputers. This made the systems economically practical for a much wider range of companies in general industry. This larger potential market for such systems spurred the development of minicomputer-based turnkey systems with all the hardware and software included in an off-the-shelf package.

Initially, these interactive graphics systems performed little more than simple automated drafting. But as computer hardware became more powerful and software was refined, the capabilities of graphics systems expanded dramatically. These systems now permit the user to perform a much wider range of geometric manipulations and sophisticated analysis. Moreover, the systems have extensive command menus that make them faster and easier to use than earlier systems.

RANGE OF APPLICATIONS

Interactive computer graphics dramatically increases design and drafting productivity. A user of a computer graphics system can generally perform routine tasks faster and more accurately than would otherwise be possible with the traditional pencil-and-paper methods.

One of the first industries to make wide-scale use of computer graphics was electronics. By the late 1960s, circuit complexity had increased dramatically, along with the sheer number of these circuits. In addition, experienced circuit designers were scarce. Computer graphics was used to permit the available manpower to meet the huge circuit-design demand. These systems proved to be highly successful, and today many circuits are designed with computer graphics systems. Figure 1.8 shows a circuit designer using standard symbols for electronic components to construct a circuit on an interactive graphics terminal. Integrated circuits and printed-circuit boards especially are designed almost exclusively with computer graphics because of their complexity. Manual design of these highly complex circuits is no longer economically justified.

Figure 1.8. Designer develops an electronic circuit using an Imlac Dynagraphic terminal. Standard symbols for components such as transistors, resistors, capacitors, and diodes permit rapid construction of complex circuits. (Courtesy of Imlac Corp., Needham, Massachusetts.)

With the success of computer graphics firmly established by the electronics industry, its use proliferated to other applications. Like circuit design, these predominantly two-dimensional applications took advantage of the computer memory to store vast amounts of data regarding some sort of physical layout. Oil and chemical companies, for example, use interactive graphics to design complex piping layouts for refineries and petrochemical plants. Civil engineers use computer graphics to design bridges, highways, dams, and other construction projects. Also, many municipalities use computer graphics as a convenient means of quickly and accurately pinpointing the precise location of public facilities such as water mains, electric cables, telephone lines, and natural gas pipelines. Some cities even use computer graphics to monitor and control traffic flow. Mapping is another application taking advantage of computer graphics to precisely define the location of mineral and oil deposits. The military is also said to be using computer graphics to display movements of equipment and troops as an aid in strategic and tactical planning.

The use of interactive computer graphics has increased rapidly in these and other fields. But the most explosive proliferation has been in the area of mechanical design and manufacturing, or CAD/CAM. This generally requires solutions of three-dimensional problems rather than the simpler two-dimensional tasks of circuit design and similar applications. Development of software to meet these more complex applications has been one of the greatest challenges in CAD/CAM. And work is continuing to refine systems. However, CAD/CAM unquestionably holds the greatest potential for future expanded use of computer graphics.

Interacting with the computer via keyboard and lightpen or other pencil-like devices, the designer specifies points and lines on the display screen to quickly construct a display-screen drawing (or model). Technically, the so-called model is actually the unseen representation of the diagram stored in the computer data base. However, the use of computer graphics to represent this data is so widespread that the term *model* is now almost synonymous with the graphical display itself on the CRT screen.

Assisting the designer in constructing a model are thousands of software aids that automate many of the tedious tasks consuming so much time in traditional manual methods. With a stroke of the pen or the push of a button, the user can move, magnify, rotate, flip, copy, or otherwise manipulate the entire design or any part of it. For example, at the push of a button the user may issue a FLIP command to produce a mirror image of a display for modeling symmetric parts. Or a TRANSLATE command may be used to create models for linear parts. Here, a cross-section is defined and then translated linearly to create a surfaced model such as the beam shown in Figure 1.9. In a similar fashion, circular parts can be easily modeled with a ROTATE command in which a cross-section is rotated about a central axis to model parts such as the circular channel in Figure 1.10. Such features take advantage of the computer's ability to repeat detailed operations quickly and flawlessly.

Other simple keyboard commands further speed the creation of 3D computer models. In working with a complex model, the user can temporarily "erase" portions of it from the screen to see the area under con-

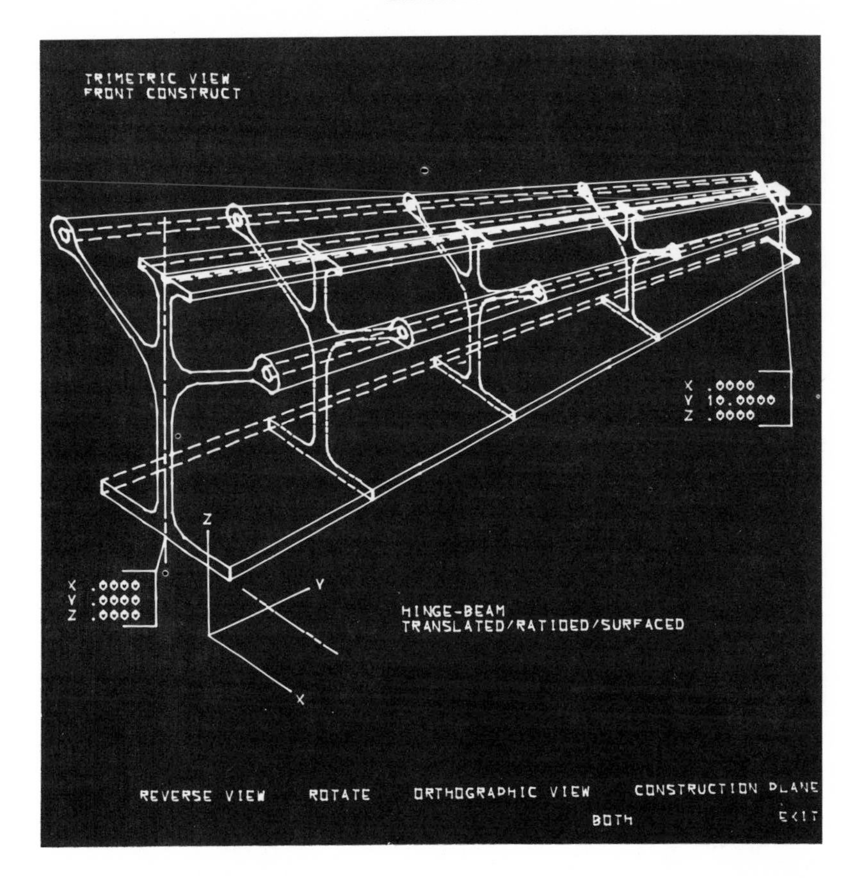

Figure 1.9. Surfaced model of a contoured beam is created by MCAUTO's CADD system by defining a cross-section line and then translating it through space with a simple keyboard command TRANSLATE. (Courtesy of McDonnell Douglas Automation Co., St. Louis, Missouri.)

struction more clearly. Then the deleted area is recalled later to complete the model. Likewise, portions of the model may be enlarged to view and add minute details accurately. And the model may be moved and rotated on the screen for the user to view at any angle. Furthermore, mechanisms such as linkages and gears in the assembly may be animated on the screen to ensure proper operation and check for interferences. When the design is

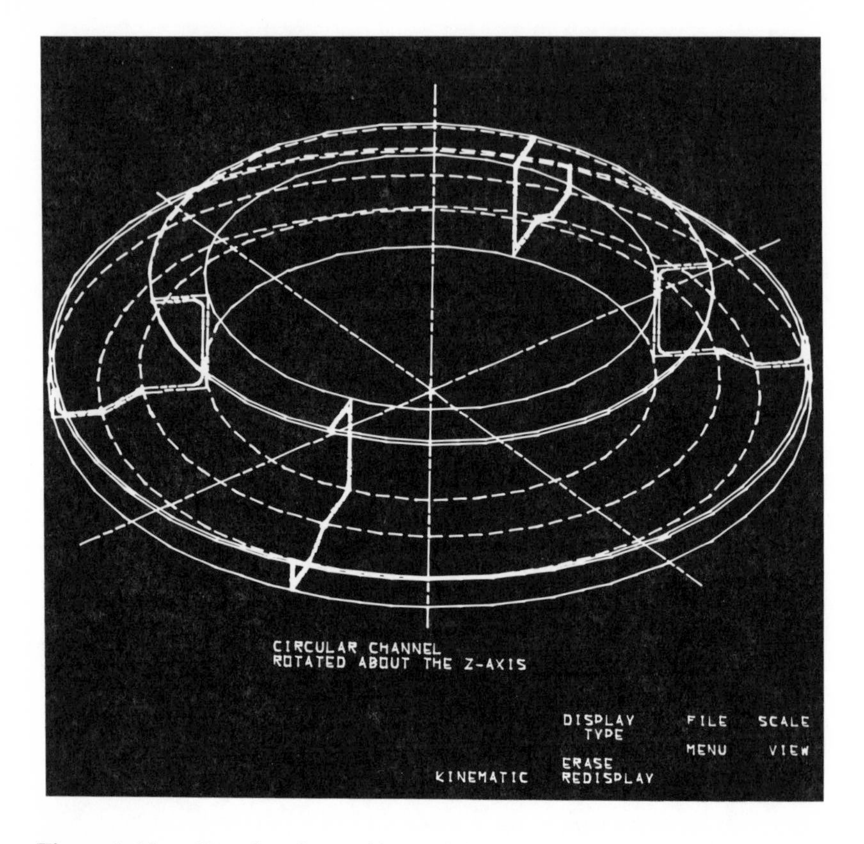

Figure 1.10. Circular channel is modeled with CADD system by defining a cross-section line and using the ROTATE command to create a surface of revolution. (Courtesy of McDonnell Douglas Automation Co., St. Louis, Missouri.)

complete, the system may automatically add dimensions and labels. And automated drafting features may be used to produce detailed engineering drawings.

After the part geometry is defined with a complete model, the user can have the computer calculate properties such as weight, volume, surface area, moment of inertia, or center of gravity. Or the powerful finite-element method may be used to determine the stress, deflections, and other structural characteristics. After the analysis, the display screen may show color-coded stress plots, the deflected shape of a part subjected to a given load, or even an animated mode shape showing how the structure might vibrate and deform during operation.

As a result, with a CAD/CAM system designers can view complex forms from various angles at the push of a button instead of having to construct costly, time-consuming physical models and mockups. Changes can be made quickly and inexpensively at the keyboard or electronic data tablet without requiring alteration of drawings or physical models. In addition, computer displays can produce realistic simulations of product operation before any hardware is produced.

After design is complete, the resulting geometric data stored in computer memory may be used to produce numerical control instructions for making the part on automated machine tools. Formerly, the preparation of NC instructions was performed manually by experienced programmers. The program was then tested on the machine and refined several times before the part was machined properly. Many of these tedious and costly operations are now reduced with CAD/CAM systems. Sophisticated CAD/CAM systems can now produce NC instructions automatically for a range of part types. And tool paths may be simulated on the display screen to verify and refine the program more quickly.

CAD/CAM is used extensively to design and manufacture automobiles and aircraft. And computer systems increasingly are used in general industry to design and manufacture bottles, camera parts, construction equipment, agricultural machinery, fasteners, electric motors, home appliances, and an ever-expanding list of products.

BENEFITS

The most obvious benefit of CAD/CAM is increased engineering productivity. This is the single consideration that influences most potential users to invest the large amount of capital necessary to implement CAD/CAM. The purchase decision is based on a trade-off of high initial cost of a CAD/CAM system versus reduced engineering costs over the life of the system.

In most cases there is an initial drop in productivity the first few months as the operators learn to use the CAD/CAM system. However, the overall productivity increase during the first year of operation is typically 2 to 1. Succeeding years may show further productivity increases as high as 20 to 1 depending on the application. However, a 3 to 1 or 4 to 1 increase is a common rule-of-thumb for most well-established CAD/CAM systems. In a typical mechanical design application, a 2 to 1 productivity increase is usually sufficient to justify the purchase of a CAD/CAM system. With this increase in productivity, the system generally pays for itself in one or two years.

Another benefit of CAD/CAM is the increased analytical capability placed at the fingertips of the user. This permits rigorous product analysis that would be virtually impossible to perform manually. In finite-element analysis, for example, the simultaneous equations describing the model typically number in the hundreds or thousands; so a computer and specialized data-handling programs are required. Moreover, computer-assisted modeling aids are usually needed to analyze most structures at reasonable time and cost. Performed manually, the method is too tedious and time-consuming to be practical.

One such modeling aid is demonstrated in Figure 1.11. The SUPERTAB package is shown in use for the analysis of a high-speed centrifugal impeller. Not only does the package aid the user in constructing the model, but after the analysis the results are displayed graphically for easy interpretation.

Another benefit of CAD/CAM in many applications is reduced product and development cost. This is a direct result of increased engineering productivity and the use of the sophisticated analysis capabilities in the CAD/CAM system. In many applications, computer simulation of an entire mechanical system or product is possible. In this manner, functional characteristics such as vibration, noise, service life, weight distribution, and stress can be analyzed with the computer instead of building costly prototypes. With CAD/CAM, computer simulation can be used to determine these characteristics early so that components can be designed accordingly. Required changes can be made and the product resimulated until a sufficiently refined design is reached. This is in contrast to the traditional build-and-test approach, where component designs are initially based on load estimates and refined with prototype testing.

For example, system models are exercised in the computer to predict how a structure will bend, twist, and rock during operation. This is a technique known as modal analysis in which animated models depict how a structure will vibrate at various modes or natural frequencies. Since the actual displacements are comparatively small, distortion is highly exaggerated in the animated mode shapes. The model is modified in an iterative process until structural performance is satisfactory.

In the development of a truck, exercising the model revealed a large cab pitch at 6.5 Hz as shown in Figure 1.12. This displacement was reduced

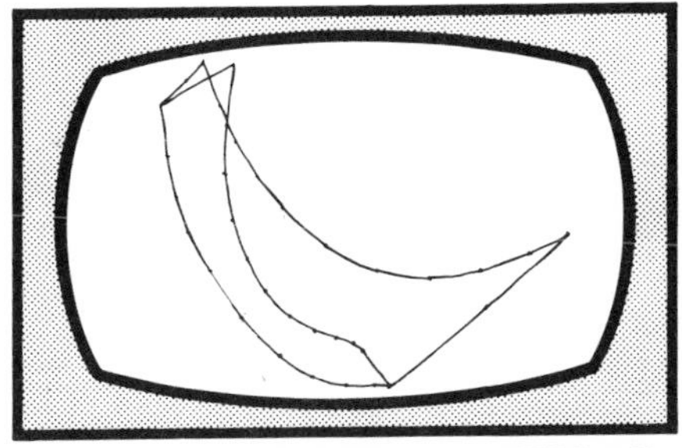

1. The geometric modeling capabilities of SUPERTAB enable the user to define the boundaries of a blade/hub section of a high-speed centrifugal impellar.

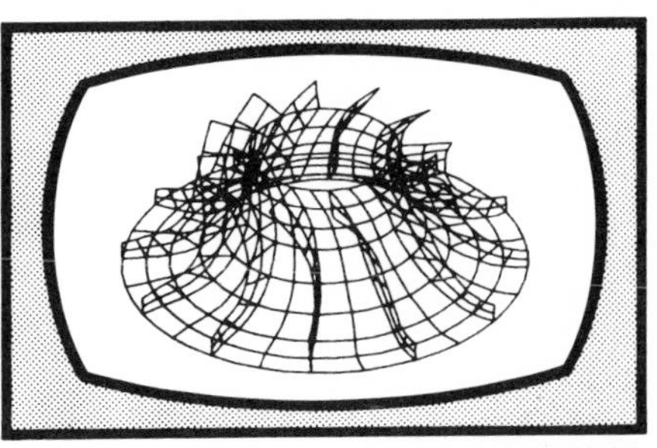

2. Once a finite element mesh is generated within the blade/hub boundaries, the section is copied eleven times to complete the model.

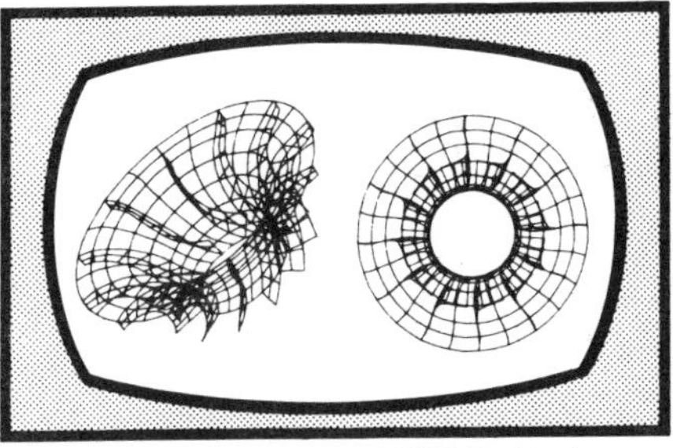

3. Extensive viewing capabilities of SDRC Graphics increase user understanding of the model and help check for accuracy.

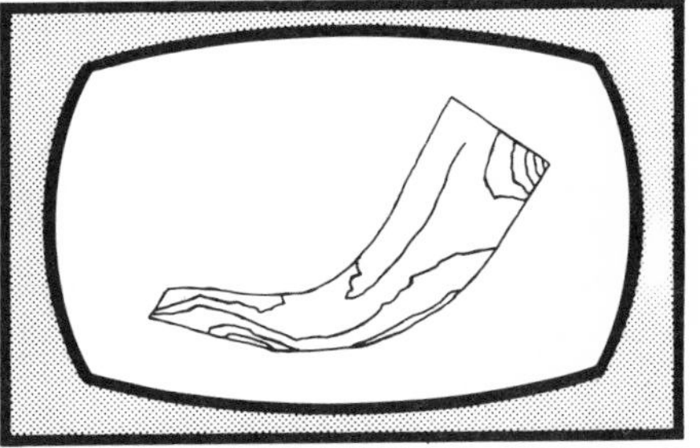

4. Displaying von mises shell stresses of the blade section, using OUTPUT DISPLAY, enables rapid interpretation of critical areas.

Figure 1.11. SUPERTAB modeling package aids user in constructing a finite-element model and displays the results graphically for easy interpretation. (Courtesy of Structural Dynamics Research Corp., Milford, Ohio.)

significantly by increasing the stiffness of the cab mounts and frame side rails. In similar fashion, the center portion of an airframe section was modified to reduce vibration levels as revealed by the displacement plots. In other words, CAD/CAM cycles through design iterations in the computer rather than on the shop floor with hardware and prototypes. In this way, CAD/CAM reduces the time and cost of developing a product. But perhaps more important, CAD/CAM designs are often closer to an optimum because the designer has time to refine it before it is committed to hardware.

Perhaps the greatest (and the most subtle) benefit of using CAD/CAM is enhanced creativity of the user. This benefit was not the initial intent of CAD/CAM, but has only recently become apparent with its increased use.

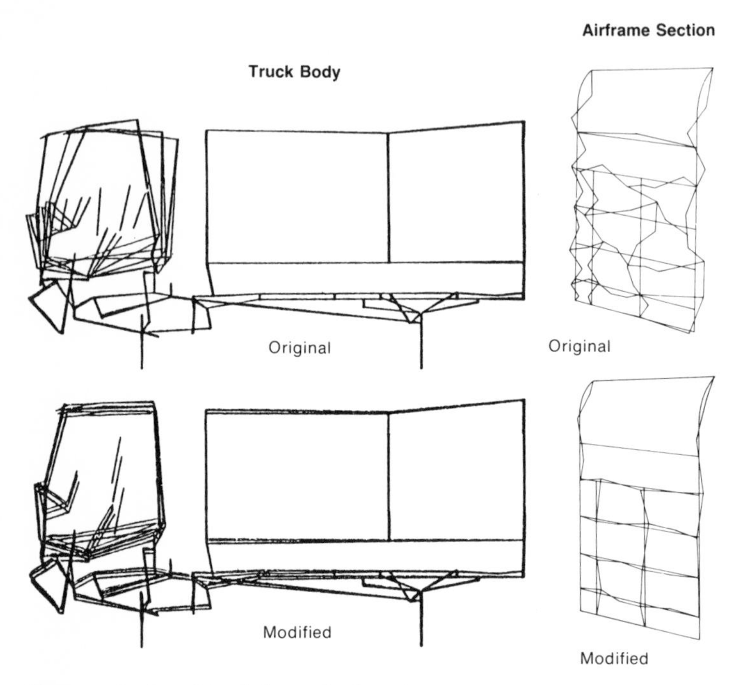

Figure 1.12. Modal analysis of a truck reveals a large cab pitch. Displacement was reduced in a modified design by increasing stiffness of cab mounts and frame siderails. (Courtesy of Structural Dynamics Research Corp., Milford, Ohio.)

Enhanced creativity with CAD/CAM is a direct result of a compatability between the human mind and interactive graphics.

According to medical and psychological experts, the human brain is not arranged as a single mainframe computer. Rather, it is comprised of hundreds of two-dimensional arrays of interconnected parallel computers made up of brain cells. Collectively, these tiny organic computers comprise a complex, sophisticated computing system. However, they are individually slow and based on relatively simple inputs and outputs.

Because of this parallel structure, information presented as printed text or tables is absorbed relatively slowly. This is because the brain must progress serially through long strings of letters, numbers, and words before it comprehends the total concept the entire assembly of information represents. In contrast, the brain absorbs the concept behind a picture or graph at a glance because the data is presented as a whole rather than in discrete parts. The ability of the brain to grasp graphical data quickly is known as *perattentive perception* and is the basis for enhanced creativity in CAD/CAM.

Basically, the unusually rapid exchange of information between the human brain and the computer enables the operator to use the CAD/CAM system as an active partner in creating new designs. This concept contrasts with the early and still prevalent concept of a CAD/CAM being an electronic slave performing only menial tasks at the direction of its user. The software behind CAD/CAM systems has now progressed to the point where the computer can actually contribute its own input to the design process.

One of the best examples of this capability is the KINSYN kinematics program. Here, the user sketches the required output motion on the display screen, and the computer responds immediately with an animated linkage mechanism to produce that particular motion. According to experts, future engineers will have programs and computers available to completely snythesize the most complex mechanical systems based solely on its intended function. In other words, the user will describe the requirements for a product, and the computer will provide a possible design to meet that set of specifications.

GROWTH OF THE INDUSTRY

In the early 1970s there were only about 200 CAD/CAM workstations installed in the world. And almost all were in-house systems in large aerospace and automotive companies that could afford them. Virtually no one else had even heard of CAD/CAM.

With the advent of the minicomputer in the mid-1970s, the cost of these systems decreased dramatically and the use of CAD/CAM took off. Since then, sales have quadrupled. Presently there are more than 12,000 workstations in place with an estimated value of over a billion dollars. Almost a half billion dollars of this equipment was shipped in 1981 alone.

CAD/CAM ranks among the fastest growing sectors of the economy. And the rapid growth shows no signs of subsiding. On the contrary, market experts say the demand for CAD/CAM systems will continue to grow through the 1980s, leading to a $2.2 billion industry by the middle of the decade. What's even more astounding are estimates that so far only 5% of the potential CAD/CAM market has been penetrated as of 1980.

Realizing the massive potential benefits of CAD/CAM, some user companies are committed to growing involvement in the technology and are investing heavily in it. One of the largest of these companies is General Electric Co.

General Electric is presently the biggest single corporate user of interactive graphics equipment and the largest customer of such turnkey vendors as Applicon and Computervision. Of a total GE capital investment of $1.4 billion in 1980, about 30% will go toward equipment and systems incorporating a significant content of CAD/CAM capability. The use of CAD/CAM at GE is growing at a rate of 20% per year. Presently, one designer in ten at GE uses an interactive graphics terminal. And by the end of the decade, corporate planners expect a tenfold increase in the number of terminals so that interactive graphics will be the standard method of design.

2
System Hardware

Most of the equipment of a CAD/CAM system is grouped in an arrangement called a workstation. This is the physical cockpit where the user controls and manipulates the vast amount of unseen data and software residing in the computer. The workstation places at the user's fingertips the power to create new designs, analyze and refine them, evaluate their performance, document their dimensions, and set in motion the manufacturing processes to make and assemble the parts.

The central item in a workstation is the graphics terminal. This has a TV-like display that may be either refresh, raster, or storage screen depending on the system. The user "draws" on this screen by specifying points and lines with a point-and-position device. This may be a lightpen touched directly on the screen or an electronic stylus moved on a sensitized tablet separate from the display. There are also electromechanical devices that rely on the user's sense of touch.

A keyboard is also an integral part of the graphics terminal. This is a typewriter-like alphanumeric arrangement for entering text and data character by character. Most systems also have a pushbutton-type function menu on which the user issues specific commands by pressing a single key. One of the most recent developments is voice-recognition in which menu commands to the system are entered verbally into a microphone.

The typical workstation also includes plotting equipment for producing hard-copy documentation. This may be a pen plotter for producing high-quality drawings or an electrostatic type for making quick review prints.

The computer system to which the workstation is connected may be comprised of a compact minicomputer, a powerful mainframe, or some combination of both. There are many ways of configuring the computing system. However, most users elect to use a minicomputer-based turnkey system purchased ready-to-use from an equipment vendor.

GRAPHICS TERMINALS

The interactive graphics terminal is the window through which the user views graphics data stored and manipulated in the computer. Seated at this terminal, the user can communicate with the computer almost instantaneously in pictures. This makes the interaction between man and computer almost conversational.

Most interactive graphics terminals display multiple views of a de-

Figure 2.1. Intelligent terminals such as this Imlac Dynagraphic unit perform many routine data-handling functions, leaving the host computer in the CAD system free for more complex analytical tasks. (Courtesy of Imlac Corp., Needham, Massachusetts.)

Figure 2.2. User verifies a drawing on a Nicolet CAD System 80. The entire drawing may be viewed on one screen while the other displays a close-up view magnified for more detailed work. (Courtesy of Nicolet CAD Corp., Concord, California.)

sign—typically front, top, and side orthographic views in combination with a 3D isometric view. These views generally are displayed simultaneously on a split screen. And design changes made on one view usually are added to the others automatically.

There are many ways in which the terminals may be configured in the workstation. Some so-called intelligent terminals like the one shown in Figure 2.1 form virtually a complete workstation in themselves. These terminals have built-in software that removes some of the data handling burden from the host computer to which they are connected. Other workstations have sets of terminals. In the system shown in Figure 2.2, for example, one screen may be used for viewing the entire drawing while another displays a magnified view for more detailed work. The dual workstation shown in Figure 2.3 permits two operators to use the system simultaneously, each working on separate projects. Still another configuration in Figure 2.4 shows a system in which one screen displays a drawing while another presents prompting instructions telling the user what to do next.

No matter what type of terminal configuration is used in CAD/CAM systems, all screens are cathode ray tubes (CRTs) that produce pictures much the same as a home television. Much work is going on to develop screens that are more compact and produce higher-quality images.

Figure 2.3. A designer at a Gerber IDS-80 dual workstation produces a wire-frame geometric model while another creates a finite-element model. (Courtesy of Gerber Systems Technology, Inc., South Windsor, Connecticut.)

However, none can rival the CRT for overall performance at this time.

The functional elements of a CRT are contained in a glass enclosure resembling a TV picture tube. A heated cathode in the CRT emits electrons that are accelerated and focused into a fine beam. This beam is deflected onto a phosphor-coated screen that glows, producing a visible trace where the beam strikes it. These high-speed traces are scanned (or rewritten) many times a second to form the images seen by the user.

There are three basic types of screens in CAD/CAM, each using a different approach for deflecting the beam and rewriting the image. These are the refresh, raster, and storage display.

Refresh screens (also referred to as vector-refresh, directed-beam,

Figure 2.4. Designer creates a drawing with a pen and data tablet on a Calcomp IGS 500 system. Right screen displays the drawing while the left screen presents prompting instructions telling the user what to do next. (Courtesy of California Computer Products Inc., Anaheim, California.)

stroke-writing, or random scan) use the beam to directly trace out the lines of the image, painting and repainting each line from endpoint to endpoint.

Because the picture is constantly being "refreshed," the image on the screen can be animated. This is an advantage in applications such as kinematics or modal analysis that often require display motion. In addition, the display can be modified, rotated, or translated on the screen without the system having to redraw the picture as with other methods. The major limitation of refresh terminals is that complex images with many line segments may appear to flicker. Typically, the flicker-free limit is about 170 inches of lines on a 19-inch screen. This is a result of the relatively long time required for the system to retrace all the lines. Another disadvantage is the large amount of random-access memory required to store an image while it is being displayed.

Raster screens (or raster-scan) use an approach similar to commercial televisions for creating an image. Instead of scanning point to point as do refresh screens, raster displays create an image with a matrix of tiny dots called pixels. The electron beam scans the entire screen from top to bottom, illuminating each pixel in an on-off pattern stored in computer memory.

Since the display is rewritten constantly, images may be animated and manipulated on the screen in real time as with refresh terminals. But because the raster scanning rate is fixed, complex images do not flicker. In addition, raster displays are bright and may display color. This feature is useful in creating stress plots, mode shapes, complex geometric models, and other images where particular details must be enhanced or otherwise differentiated. The major limitation of raster displays is poor line quality. Vertical and horizontal lines are acceptable, but the discrete nature of the matrix dots may make diagonal lines appear jagged.

The storage display (or direct-view storage tube) initially creates an image with a directed beam of electrons that traces the picture line-for-line in a manner similar to refresh displays. But unlike refresh screens, the storage tube does not continuously retrace the image lines. Rather, a flood gun constantly bombards the entire screen with electrons that by themselves have energy just below the threshold to illuminate the phosphor. When struck by the high-energy writing beam, the screen changes potential in the vicinity of impact, allowing the flood-gun electrons to illuminate the phosphor along the track indefinitely.

Storage tubes have excellent resolution and can display extremely complex images with relatively little computer memory and processing. The major limitation of storage tubes is that no selective erasing is possible. Once an image is displayed and stored, it must be completely erased and repainted on the screen from scratch to change any part of it. In a complex graphical display, repainting the entire image to change a small detail can be time-consuming. For this reason, image manipulation and animation are not possible with the storage tube. In addition, relatively low brightness and contrast levels often require ambient lighting to be dimmed or otherwise screened from the display.

POINT-AND-POSITION DEVICES

The user of a CAD/CAM system creates geometric constructions on the CRT with hand/eye coordinated motions similar to the kind used in drawing with pencil and paper. He may simply point to a location on the screen and enter a command to add or erase a point or line. He may sketch an entire drawing and have the system straighten the lines and add dimensions. Or he may circle a portion of the display and move it, magnify it, or erase it.

A variety of input devices are available for such pointing and positioning tasks. These include lightpens, electronic tablets, and a variety of electromechanical devices.

The lightpen interacts directly with the image on the CRT screen as shown in Figure 2.5. A tiny photocell mounted at the end of the pen-shaped device produces an electronic signal when light is detected. The signal is routed to the display processor, which locates the point on the screen by determining which display element was being illuminated when the signal was generated. Lightpens are used in many CAD/CAM systems because their apparent direct contact with the display image makes them convenient input devices. However, the tracking program that identifies pen position on the screen involves additional software and processing. And lightpens only work with continuously rewritten images on refresh and raster screens.

Other input devices use an indirect approach in which the input device is manipulated at a distance from the screen. Some systems use a touch-sensitive electronic tablet and stylus like the arrangement shown in Figure 2.6. This is generally a flat, sketch-pad arrangement that may be physically part of the terminal or separate from it. The user maneuvers the stylus on the tablet, which is electronically sensitized to detect the position of the stylus. The stylus is used like a pencil in locating points and specifying positions to create geometric constructions on the screen. Visual feedback is provided to the user by a set of cross-hairs on the screen.

Other systems use free-moving, hand-held cursors like the one in Figure 2.7. These puck-like devices are used primarily for entering existing drawings and sketches into the CAD/CAM system. In this process (called digitizing), the cursor is placed over a point on the drawing and a key is pressed to enter the position. In this manner, the user enters all the points and connectivity commands for the lines that make up the complete drawing. Digitizing the points enters an idealized form of the sketch into the computer, which cleans up the drawing by performing such tasks as straightening line segments, smoothing curves and arcs, and orienting lines at proper angles to one another. Since cursors are used mostly for digitizing large drawings, the tablets typically are drawingboard-sized units. Some of these larger tablets have cursors constrained in a moving gantry framework. This type of digitizer sometimes is fitted with a penholder and is a hybrid device called a digitizer/plotter.

Many electromechanical devices relying strongly on the operator's sense of touch are available to manipulate cross-hairs on the screen. An arrange-

Figure 2.5. MCAUTO's CADD system uses a combination of light pen, typewriter keyboard, and a smaller menu keyboard to create a model. (Courtesy of McDonnell Douglas Automation Co., St. Louis, Missouri.)

Figure 2.6. The Applicon IMAGE system uses an electronic tablet and stylus to manipulate points and lines and to select specific menu commands. (Courtesy of Applicon Inc., Burlington, Massachusetts.)

Figure 2.7. Points and lines are entered into Summagraphics Datagrid II computer-aided drafting system with a hand-held digitizing cursor positioned on a rough sketch. (Courtesy of Summagraphics Corp., Fairfield, Connecticut.)

ment of two thumbwheels for the x and y directions, and a third-dimensional capability, may be added by moving the stick in and out or twisting it clockwise or counterclockwise. Another electromechanical arrangement uses a trackball consisting of a 4-inch-diameter ball mounted in a bearing so that it spins freely in all directions. The direction of cross-hair movement on the screen is controlled by the ball's rotation. Other systems use a set of four pushbuttons that move the cross-hairs up, down, right, or left. A fifth button is sometimes included to return the cross-hairs to an initial position or to control their speed.

KEYBOARDS

The typewriter-like alphanumeric keyboard is used to issue instructions and enter data into the CAD/CAM systems. The keyboard is used to enter text onto drawings, specify coordinate positions on the screen, store and retrieve data from the computer, request status information, send messages to other workstations, and many other tasks. The keyboard performs a wide variety of functions in the CAD/CAM system and is the systems universal input device.

Data is entered on the alphanumeric keyboard character-by-character. Even simple commands require a large number of keystrokes, making the process tedious and error-prone for long command sequences. Thus, most CAD/CAM systems have function menus with which frequently used commands can be issued merely by pressing a single key on an assembly separate from the main alphanumeric keyboard. Figure 2.8 shows an operator using an alphanumeric keyboard, above which the function

Figure 2.8. Keyboards are common to all computer terminals for entering instuctions and data. In addition, many systems also have function menus where one key initiates a particular programmed routine. The menu here is on the slanted panel above the main alphanumeric keyboard. (Courtesy of Auto-Trol Technology Corp., Denver, Colorado.)

menu is presented on a slanted panel. Typical functions in a geometric construction menu are DEFINE A POINT, TRIM A LINE, CONSTRUCT A CONIC, DELETE A COMPONENT, and ROTATE A COMPONENT.

Function menus are generally presented as an array of pushbuttons. When a system has no more than twelve menu items, they are usually assigned to large electromechanical buttons mounted on a table-top housing placed close to the main keyboard. Systems with a large number of function menu items (sometimes over 200 items) use a more compact array of squares on a pressure-sensitive panel adjacent to the main keyboard.

The types of menu items vary from application to application. Therefore, the menus of most CAD/CAM systems can be readily changed by loading an appropriate operating program into the system. Overlay sheets on the function menu then identify the key functions. In this way, the same CAD/CAM system can be used to design, for example, mechanical structures as well as integrated circuits. In addition, some systems also permit the user to enter very specific menu items used frequently only in narrow applications. Typically these are the specialized symbols or groups of symbols required to produce drawings. For example, geometries of bearings and cutting elements were added to the standard menu of drawing commands in an Applicon system used by a company exclusively to design oil-well rock-drilling bits.

In some lightpen systems, the function menu is displayed on the CRT screen. Menu items are selected by pointing at the individual location with the pen. These so-called *light buttons* may be displayed continuously at a reserved location on the screen, or they may be called up by the operator when needed. Light-button menu items may be an array of symbols representing the functions or complete text statements describing the functions.

One of the most recent innovations in function menu selection is voice data entry. In this method, the operator enters menu commands verbally by speaking into a microphone as in Figure 2.9. These systems have the capacity to store almost 100 menu items (called a vocabulary) created by the user. The user "trains" the system to recognize the vocabulary by entering the menu function into the keyboard and speaking the menu command word into the microphone several times. Because of differences in voice inflection, accent, and enunciation, each operator must individually train the system.

Voice data entry is a very efficient method of selecting menu items, since it frees the operator's hands to manipulate other devices and permits him

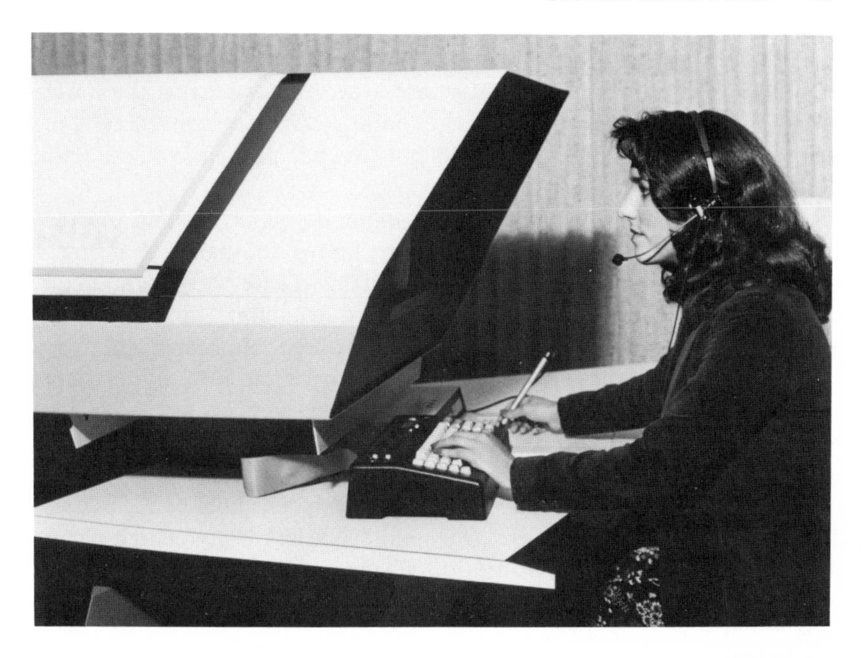

Figure 2.9. Voice control unit on the Calma GDSII system permits the user to enter verbal menu commands such as CONTINUOUS PAN, ZOOM, MULTIPLE VIEW, and COMPOSITE IMAGE. (Courtesy of Calma Co., Santa Clara, California.)

to concentrate more attentively on the display screen. However, voice changes caused by laryngitis or other conditions may prevent the system from recognizing a command. And infrequent users may have difficulty remembering the vocabulary.

PLOTTERS

An integral part of the CAD/CAM workstation is the plotter, the device that converts the digital graphics data from the CAD/CAM system into hard copy. Plotters generally produce detailed design engineering drawings, assembly diagrams, schematics, logic diagrams, flow charts, illustrations, and a wide range of other output documentation.

Specialized plotters are available to meet specific requirements in the CAD/CAM system. For example, photoplotters are high-precision devices used mainly for the production of printed-circuit artwork masters. And computer-output microfilm plotters are available for placing the CAD/CAM system output graphics directly on microfilm for convenient storage. However, most CAD/CAM systems use pen plotters, sometimes in combination with an electrostatic type.

Pen plotters use ink pens to draw graphics on paper, vellum, or other drawing medium. Bed-type pen plotters are known for their high accuracy and are used most often in CAD/CAM systems. These types use standard drawing paper that is rigidly attached to a bed structure. The bed may be a stationary flat bed, a rotating cylinder, or a moving belt. In drum plotters, special paper with holes on the sides is moved back and forth by sprockets on a tractor drum. Drum plotters are less expensive than bed types, and some have overall lower accuracy. For this reason, they are used for less demanding CAD/CAM drawing applications.

Electrostatic plotters produce an image on paper as lines of raster dots. Electrostatic point charges are deposited on dielectrically coated paper. The point charges become visible dots when a liquid hydrocarbon toner is applied to the paper. The principle advantage of electrostatics is their high speed, which may be up to 100 times faster than pen plotters. The main limitation of electrostatics is low accuracy. In addition, special conductive paper must be used. And electrostatic plotters tend to be more expensive than pen types. Furthermore, additional software must be included in the system for converting the standard vector output from the computer into raster format for use by the electrostatic plotter.

In spite of these limitations, electrostatic plotters are being used increasingly in CAD/CAM systems. Usually, they are used for their high speed to produce quick preliminary drawings. After the design is complete and refined, the pen plotter is used to produce accurate, higher-quality documentation.

COMPUTER NETWORKS

A computer consists of a central processing unit, memory, and data buses. Input data and programs telling the computer how to handle the data are stored in a memory. Program instructions are carried out and the data manipulated by a central processing unit. Digital signals flow between the memory, central processing unit, and peripheral equipment like plotters and digitizers on data buses, which act as highways on which data traffic flow is coordinated.

The simplest CAD/CAM system consists of one workstation interacting with a single computer, which is generally a minicomputer. However, most systems have several workstations connected to the minicomputer. This sharing of data storage, processing, and output facilities like printers and plotters reduces the cost per terminal. Typically, three or four workstations may be used simultaneously with a single minicomputer. Additional workstations tend to saturate the computer with work and may produce noticeable delays in the computer's response to operator inputs. However, up to 16 workstations may be supported by a single minicomputer if their use is staggered.

To implement a CAD/CAM system, users may purchase the individual pieces of equipment and required software to build their own system. However, most users find it more convenient to purchase a turnkey system from a vendor such as Computervision, Applicon, Calma, Auto-trol, or Gerber. Here the supplier has total responsibility for building, installing and testing the complete system.

Complete turnkey systems generally cost $300,000. But the system typically pays for itself in increased engineering productivity in about one year. Generally, most companies supporting 15 to 20 designers can afford to purchase a turnkey system.

Some large CAD/CAM users link together two or more stand-alone systems to form a network of workstations and minicomputers all operating with the same data base and all sharing access to the same information.

In many of these larger systems, mainframe computers are used for finite-element analysis, huge matrix manipulations, and other complex processing tasks. These can be performed quickly on a large mainframe whereas they might take days on a minicomputer. The mainframe may be part of a company's in-house computer network. Or timesharing services allow the user to lease time and access the appropriate software in remote computers provided by vendors via telephone lines.

Some of these systems are arranged in a hierarchical distributed processing network. In this arrangement, processing tasks are shared by different levels of minicomputers. Processing in the entire network may be coordinated by a central mainframe.

An interconnected system also can be created with a mainframe computer and several so-called intelligent terminals. These terminals can be purchased for about $15,000, and for large systems they may be more cost-effective than linking together turnkey systems. Usually the terminals have built-in graphics software that removes some of the data-handling burden from the host computer.

3
CAD/CAM Functions

A user at a CAD graphics terminal can design a part, analyze stresses and deflections, study mechanical action, and produce engineering drawings automatically. And from the geometric description provided by CAD, production people may produce NC instructions, generate process plans, program robots, and manage plant operations with a CAM system.

These two technologies and their associated capabilities are being combined into integrated CAD/CAM systems. In other words, a design is created and the manufacturing process is controlled and executed with a single computer system. These sophisticated systems are now used in a few large manufacturing operations, but an increasing number of plants are gaining these advanced capabilities. And some experts predict that integrated CAD/CAM systems will ultimately lead to the automated factories that forward-looking managers have long envisioned.

This evolution of CAD/CAM is integrating many diverse technologies that have developed independently since the early 1950s. Initially, automated drafting stations were developed in which computer-controlled plotters produced engineering drawings. The systems later were linked to graphic display terminals where the user created geometric models describing part shape. Drawings were then produced from the resulting data base in the computer.

At a graphics terminal, the user could communicate with the computer in pictures instead of columns of raw numbers. As a result, the computer could be easily used by those untrained in programming. Now, advanced systems with interactive graphics terminals have powerful analytical programs that evaluate parts with techniques such as the finite-element method. And the complex motions of linkages and other mechanisms can be studied with advanced kinematic programs.

At the same time CAD technology was emerging, CAM advancements also were being made—mostly in numerical control. Until recently, only experienced programmers could develop and verify NC instructions. But now, some of these instructions can be generated automatically, and computer simulation can quickly verify tool paths. Some systems also have process planning features for determining the sequence of fabrication steps. And factory management capabilities may be available for directing the flow of work and materials through the plant. The newest feature of CAM is robotics, where automated manipulator arms handle tools and workpieces and perform other manufacturing tasks.

A major milestone in the computer technology was the combination of CAD features (geometric modeling, analysis, kinematics, and automated drafting) with the CAM capability of automatic NC tape preparation. This finally bridged the gap separating CAD and CAM, making it possible for an engineer to proceed from initial concept to finished part with one integrated system.

Efforts to advance and integrate all CAD/CAM functions into one unified technology are aimed at creating what many observers are calling CAE—an acronym for computer-aided engineering. In an idealized system, all CAD/CAM functions are interfaced with a common data base as shown in Figure 3.1. Presently, many systems have CAD functions integrated, and a few have limited NC capabilities. But many other CAM functions are not fully developed and are separately implemented.

The cost of CAD/CAM systems has decreased just as dramatically as capabilities have increased. As a result, what was exotic and prohibitively expensive a few years ago is now commonplace. Only ten years ago, a computer and graphics equipment for a CAD/CAM system cost several million dollars and could be afforded only by a few giant companies. Now, equivalent systems cost a few hundred thousand dollars and are within the budget of many other manufacturers.

As a result of these economic and technical developments, CAD/CAM has gradually permeated general industry. Major users of the most sophisticated systems are still very large companies, most notably in the automotive and aerospace industries. But a growing number of smaller manufacturers are using computer systems to design and fabricate a wide range of general products such as home appliances and machine tools.

Many experts point to CAD/CAM as the key antidote for sagging productivity, and they predict the future widespread use of CAD/CAM

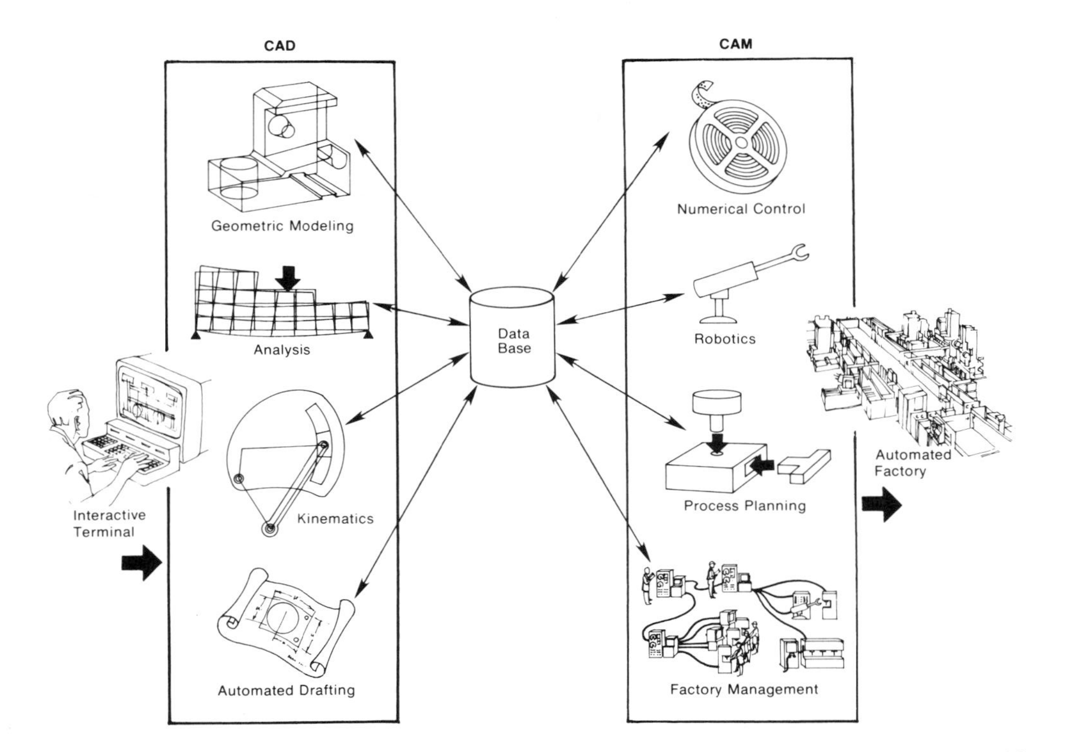

Figure 3.1. In an idealized CAD/CAM system, the user interacts with the computer via a graphics terminal, designing and controlling the manufacturing process from start to finish with information stored in a shared data base. Efforts to advance and integrate the individual segments of CAD/CAM technology are aimed at creating what many observers are calling *computer-aided engineering,* or CAE. (Courtesy of *Machine Design Magazine,* Cleveland, Ohio.)

systems throughout general industry. But these experts also acknowledge that CAD/CAM must be greatly refined to realize its full potential. This refinement is viewed as twofold. First, the technology of individual CAM functions must be advanced to the level of CAD sophistication. And, second, individual CAD/CAM functions must be combined into truly integrated systems.

Cooperative efforts to advance CAD/CAM to this level are underway on several fronts. The international CAM-I organization is developing individual areas of CAM technology. And framework programs being developed jointly by the U.S. Air Force Integrated Computer Aided Manufacturing (ICAM) project and the NASA Integrated Program for Aerospace Vehicle Design (IPAD) project are linking together individual CAD/CAM functions. Furthermore, IGES (Initial Graphics Exchange Specification) is attempting to standardize data communication between separate CAD/CAM systems.

CAD/CAM technology is changing so rapidly that it appears confusing to many people—even to experts directly involved with its development. Some regard CAD only as automated drafting and CAM as merely NC tape preparation. Others include all engineering tasks performed with a computer as CAD/CAM. Actually, CAD/CAM is comprised of distinct functional areas. Experts group CAD functions in four major categories: geometric modeling, engineering analysis, kinematics, and automated drafting. And present activity in CAM technology centers around four main areas: numerical control, process planning, robotics, and factory management.

GEOMETRIC MODELING

The designer constructs a geometric model on the CAD/CAM terminal screen to describe the shape of a structure to the computer. The computer then converts this pictorial representation into a mathematical model which it stores in a data base for later use. The model may be recalled and refined by the engineer at any point in the design process. And it may be used as an input for virtually all other CAD/CAM functions.

Because so many functions depend on the model, geometric modeling is considered by many experts to be the most important feature of CAD/CAM. For example, the geometric model may be used to create a finite-element model for stress analysis. It may serve as an input for computer-assisted drafting to produce engineering drawings. Or it can be used as a basis for producing NC tapes for fabricating the part.

Most modeling is done with wire frames that represent the part shape with interconnected line segments. Depending on the capabilities of the system, the model may be 2D, 2½D, or a full 3D model. However, even 3D wire-frames often do not adequately represent the solid nature of an object and sometimes require further definition by the user. An advanced geometric modeling technique that overcomes this problem is 3D solid modeling.

In the most common form of 3D solid modeling, models are constructed with building blocks of elementary solid shapes called primitives. One advanced system is the SYNTHAVISION program, a software package developed by Mathematical Applications Group Inc. Figure 3.2 shows this software package being used to create a 3D model of a part. Another commercially available program is PATRAN-G developed by PDA Engineer-

Figure 3.2. The SYNTHAVISION program enables engineers to create geometric models in high-resolution color and 3D perspective. On this system, the model can be rotated, viewed in cross-section, or exploded for detailed study. (Courtesy of Control Data Corp., Minneapolis, Minnesota.)

ing. This program uses interconnected surfaces to define the boundaries of the solid and has processing features for readily transforming the geometric model into a finite-element model. There are now over 12 such commercially available programs. Other experimental solid-modeling programs have been developed around the world, mostly in universities. However, these generally are not as refined and lack the extensive software support of the commercially available programs.

Because the geometric model ultimately is used as a basis for machining a part, geometric modeling is tied closely to NC technology. A pioneer in the move to tie these two areas together is the CAM-I organization, which is studying solid modeling schemes to establish a common link between geometric models and NC systems.

CAM-I also has been the driving force in the development of an advanced system to define and create NC tapes for sculptured surfaces. These are complex contours such as pop bottles, camera cases, turbine blades, and automobile bodies that cannot be described with the usual lines and curves of conventional modeling systems.

The technologies of 3D solid modeling and sculptured surfaces processing are two of the most recent advancements in geometric modeling. And these areas continue to be refined and integrated with NC features. Eventually experts predict that integrated systems will be able to model and machine any general shape.

ANALYSIS

Most CAD/CAM systems permit the user to move directly from the geometric model to analytical functions. For example, simple keyboard instructions may command the computer to calculate weight, volume, surface area, moment of inertia, or center of gravity of a part. Some of these mass property calculations for a typical part are shown in Figure 3.3.

The most powerful method of analyzing a structure on a computer is probably the finite-element method. In this technique, the structure is represented by a network of simple elements that the computer uses to determine stresses, deflections, and other structural characteristics. A finite-element model is shown for a typical part in Figure 3.4.

Manual construction of the element mesh is a tedious and time-consuming task; so computer aids are virtually indispensable in creating complex finite-element models. And the analysis requires tremendous data-manipulation capabilities only available on large computers. As a result, implementation of the finite-element method is difficult without a CAD/CAM system.

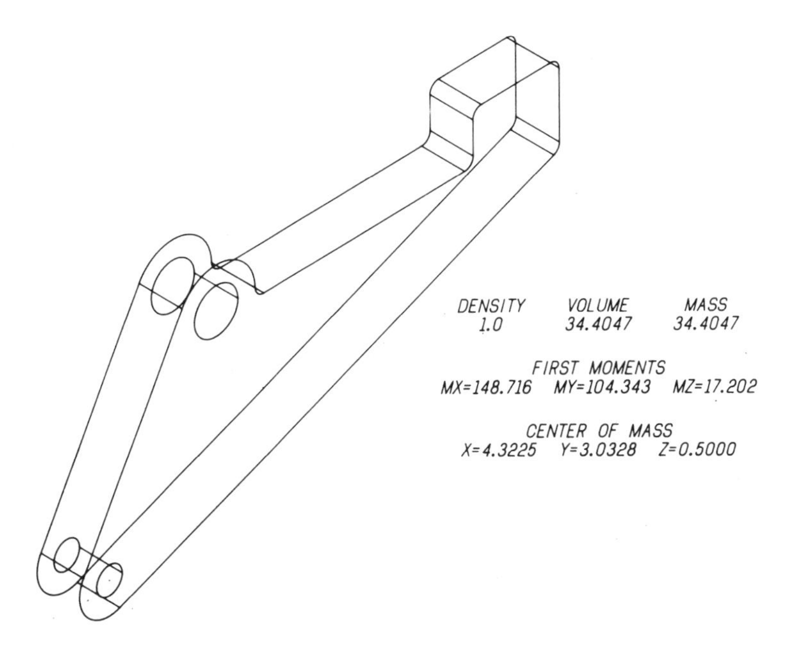

Figure 3.3. Some CAD/CAM systems perform mass property calculations such as these with simple keyboard commands. (Courtesy of Computervision Corp., Bedford, Massachusetts.)

In integrated systems, the user can call up the geometric model of the part and create a finite-element model quickly and easily using automatic node-generation and element-generation routines. Once a part is modeled, the user specifies loads and other parameters. Then the model may be analyzed with programs as NASTRAN, STRUDL, ANSYS, or SAP.

The huge amounts of output data produced from the analysis may be condensed into graphic form with post-processing features for quick evaluation. For example, a deflected shape may be superimposed over the original model to show structural deformation. Areas of high stress may be pinpointed with color-coded plots. Or animated mode shapes may depict how the structure vibrates during operation.

If the finite-element analysis reveals excessive stress or deflection, the computer model may be modified and reanalyzed. In this manner, the designer can evaluate structural behavior before the product is actually built. And the design can be appropriately modified without building costly physical models and prototypes.

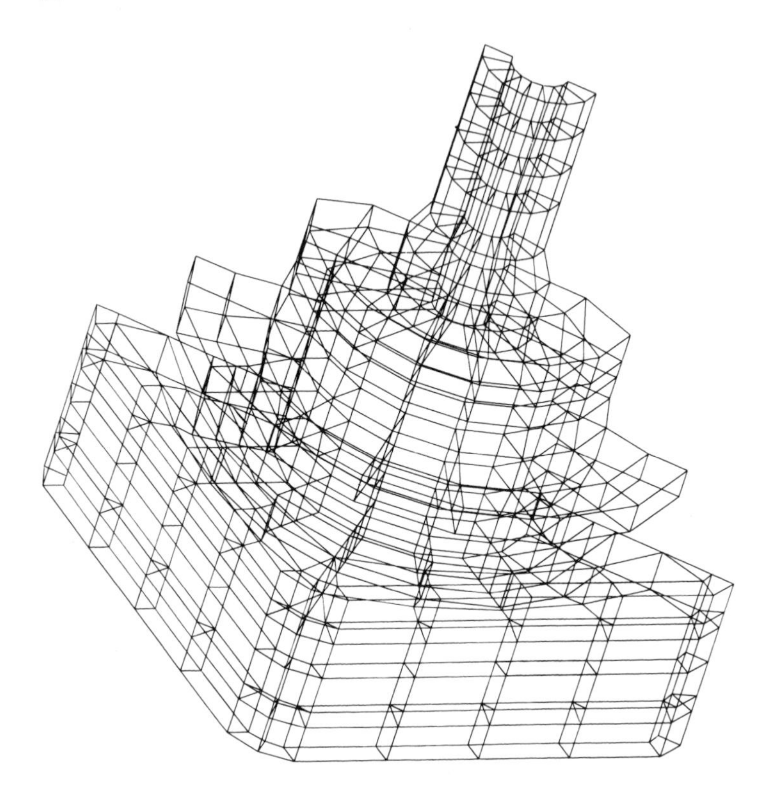

Figure 3.4. Finite-element models such as this one created with the UNISTRUC system are networks of line elements representing the structure in the computer. (Courtesy of Control Data Corp., Minneapolis, Minnesota.)

One of the most recent developments in CAD/CAM analysis concerns the combination of analytical and experimental data to create a total system model. In this technique, rigid parts are analyzed with finite-element techniques. And characteristics of elastic components such as shock absorbers and isolation mounts are determined by testing. The data is combined to create a system model, which is then exercised to predict structural behaviour during operation. For example, input data for an automotive analysis may simulate wheel unbalance, braking, turning, or tire impact with a curb or chuckhole. The computer predicts the response of the overall vehicle to these conditions. The method is useful in improving the structural integrity of a range of vehicles such as trucks, buses, and tractors. But it has also been useful for other machines such as machine tools and home appliances.

KINEMATICS

The equations associated with complex mechanisms such as four-bar linkages are extremely difficult to set up and solve. As a result, designers traditionally used pin-and-cardboard models or cumbersome graphical methods to develop practical mechanisms.

CAD/CAM kinematic programs now strip away much of this tedium, cost, and complexity from mechanism design. Most systems can plot or even animate the motion of simple hinged parts such as doors and cranks to ensure they do not impact or interfere with other components. In addition, more sophisticated programs are available for designing and analyzing the most complex mechanisms.

These kinematic programs permit the user to develop mechanisms much faster than would be possible otherwise. For example, an automobile hood linkage can be designed in only a few minutes of interaction with a computer. In contrast, hundreds of man-hours can be required to manually design the part. In most cases, the final mechanism design approaches an optimum because the computer allows so many design alternatives to be compared. For example, the computer simulation of the front-end loader in Figure 3.5 permitted the complex linkage paths to be determined and evaluated easily.

Hundreds of kinematics programs have been developed in universities to solve very specialized problems, but only a few are refined sufficiently to be commercially practical. Two prominent programs are ADAMS and DRAM, developed by Milton Chace of the University of Michigan. And IMP was created by John Uicker of the University of Wisconsin. KINSYN was developed by Roger Kaufman of George Washington University, and LINCAGES by Arthur Erdman of the University of Minnesota.

Users may obtain these refined programs in several ways. Some of them may be run on timesharing networks. But frequent users often find that purchasing the programs outright for use on in-house systems is more economical. Also, early versions of some programs are in the public domain and may be obtained relatively inexpensively. These early versions lack the refinement and support of commercial programs, but generally they can be purchased for a small documentation charge of a few hundred dollars.

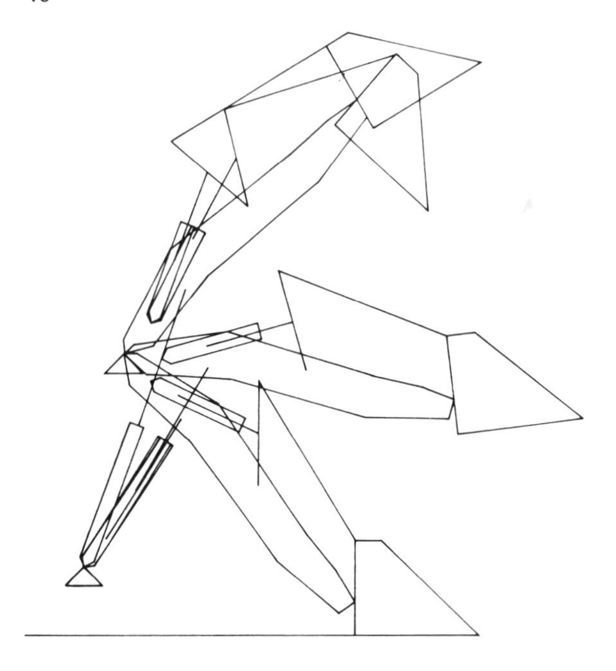

Figure 3.5. Complex mechanisms such as this front-end loader simulated by the IMP program may be designed quickly without traditional pin-and-cardboard models or lengthy mathematical equations. (Courtesy of Structural Dynamics Research Corp., Milford, Ohio.)

The ADAMS, DRAM, and IMP programs require the user to enter problem-oriented language statements as inputs. The computer then produces link positions, forces, velocities, accelerations, and other output data. In contrast, KINSYN and LINCAGES develop mechanism designs based on required motion paths. LINCAGES requires the user to enter path data on a keyboard, whereas KINSYN utilizes more interactive graphics. The user specifies motion requirements on an electronic tablet with a stylus. The computer immediately displays a linkage configuration capable of providing that motion path on the terminal screen.

The computer's capability to develop a complex mechanism design has far-reaching implications in CAD/CAM technology. Some experts envision future programs and computers that may completely synthesize a complex mechanical system based solely on its intended function. In other words, the computer will provide a design created only from sets of

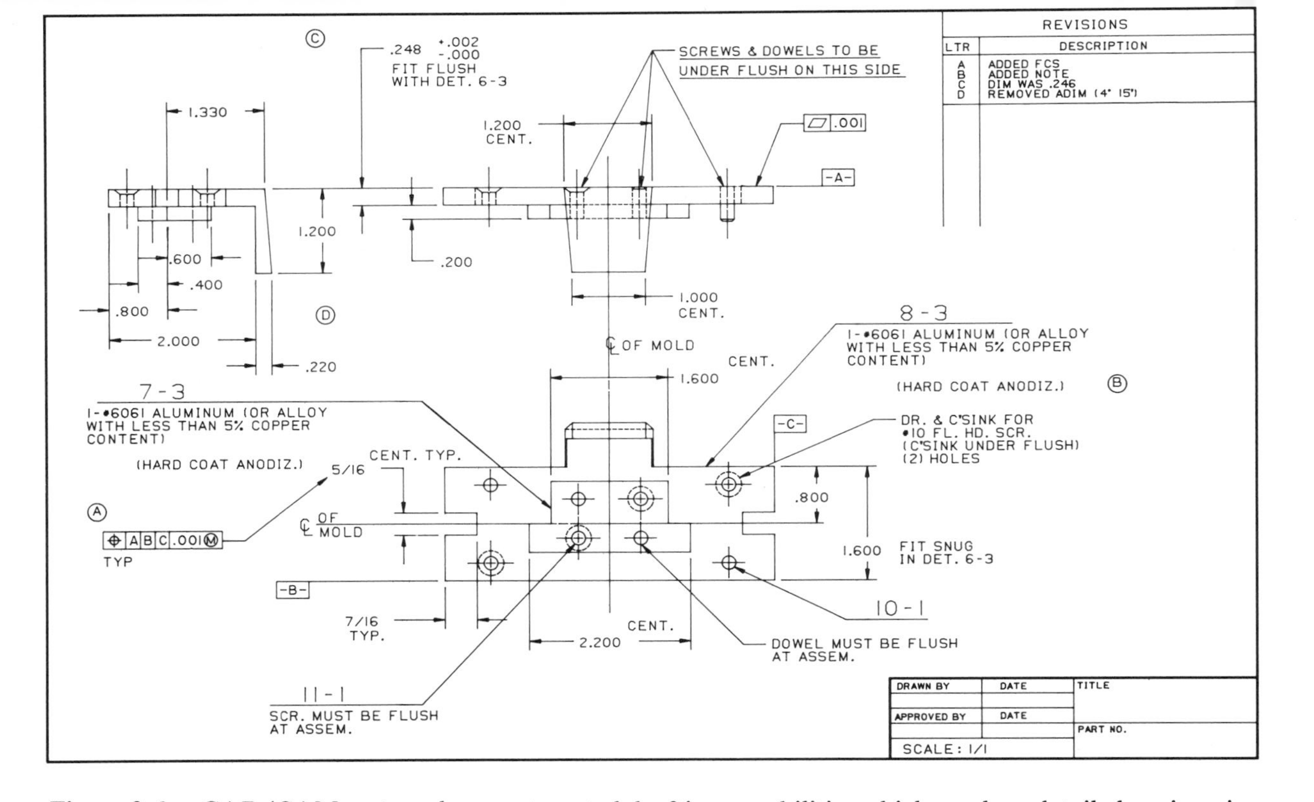

Figure 3.6. CAD/CAM systems have automated drafting capabilities which produce detailed engineering drawings much faster than traditional manual methods. (Courtesy of Computervision Corp., Bedford, Massachusetts.)

specifications described to it by the user. Some experts predict that design in the year 2000 may be done in a cockpit arrangement where both design synthesis and analysis may be performed for automobiles, aircraft, and other complex mechanical systems.

DRAFTING

Computer-assisted drafting features automatically produce detailed engineering drawings on command from the geometric-model data base or from inputs entered by the user at the graphics terminal.

These systems have many features that automate a range of drafting tasks to speed the production of drawings. Most systems have automatic scaling and dimensioning features. And changes made to one view may be automatically added to other multiple views. Moreover, function menus permit the user to specify points, locate lines, enter text, and produce cross-hatching at the push of a button. A typical drawing produced with these features is shown in Figure 3.6.

These automatic features coupled with the high speed of computer-driven plotters enable users to produce new drawings five times faster than with manual drafting methods. And design changes can be made up to 25 times faster with CAD/CAM. In the future, even greater productivity is expected as plotting speeds increase and more terminals are operated simultaneously from a central computer.

NUMERICAL CONTROL

One of the most mature areas of CAD/CAM technology is numerical control, or NC. This is the technique of controlling a machine tool with prerecorded, coded information to make a part. NC instructions usually are written either in the APT or COMPACT II language. COMPACT II is used in most installations. But APT is the original and more universal language and is considered by most experts to be the standard of the industry.

NC instructions generally are stored on punched paper tapes or magnetic tapes for controlling the machine tool. More advanced systems may use computer numerical control (CNC), a set-up in which the machine is hardwired to a minicomputer where NC instructions are stored. The most sophisticated systems use a direct numerical control (DNC) scheme in which several minicomputers are linked to a central mainframe. Some DNC systems eliminate the intervening minicomputer

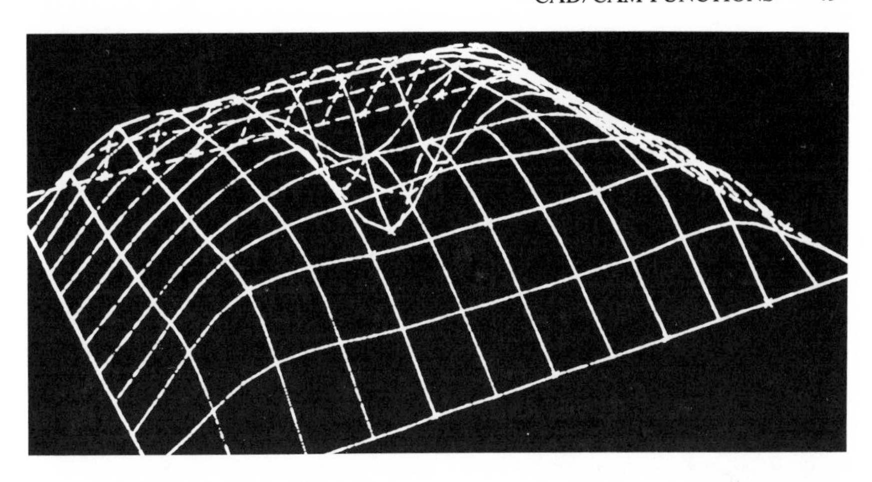

Figure 3.7. Complex contours are represented by a network of interconnected patches in the CAM-I Sculptured Surfaces Processor. After the surface is modeled, the processor output may be used to produce NC instructions to fabricate the part. (Courtesy of CAM-I Inc., Arlington, Texas.)

in favor of a direct interface between the central computer and machine tool.

Traditionally, experienced programmers write NC instructions directly from engineering drawings. Then the program is tested on the machine tool and refined several times to remove errors. These time-consuming iterations can significantly increase the cost of machining a part.

Advancements in CAD/CAM technology now make creation and verification of NC instructions much easier. For example, less machine-tool time is spent verifying the cutting paths by checking the tooling program with computer simulation. Perhaps more important, the computer itself can now generate NC instructions directly from the geometric data base. These automatic capabilities are generally restricted to flat or turned parts, highly symmetric geometries, and other specialized shapes. The Lockheed-California CADAM system marketed by IBM and the MCAUTO CADD system are examples of such systems. Future systems are expected to accommodate more generalized shapes.

Present systems must be greatly refined before NC instructions can be produced automatically for completely arbitrary geometries. A big step in this direction is the development of the CAM-I Sculptured Surfaces Processor, which models and develops machining instructions for complex contoured shapes. Figure 3.7 shows a model and machining operations demonstrating the processor on a test part. CAM-I development work on their Advanced NC Processor also brings the technology close to an integrated, generalized system that many experts claim is not far in the future. According to one CAM-I spokesman, "The problems are immense but not unsolvable. It is a matter of solving an enormous jigsaw puzzle, cleverly arranging and applying existing technology."

PROCESS PLANNING

Numerical control is concerned with controlling the operation of a single machine. However, process planning is a much broader function that considers the detailed sequence of production steps required to fabricate an assembly from start to finish. Essentially, the process plan describes workpiece status at each workstation along the line. As such, process planning has been a part of manufacturing for some time. But only recently has the computer been used in this activity.

One important aspect of process planning systems is group technology. This concept organizes similar parts into families to allow fabrication

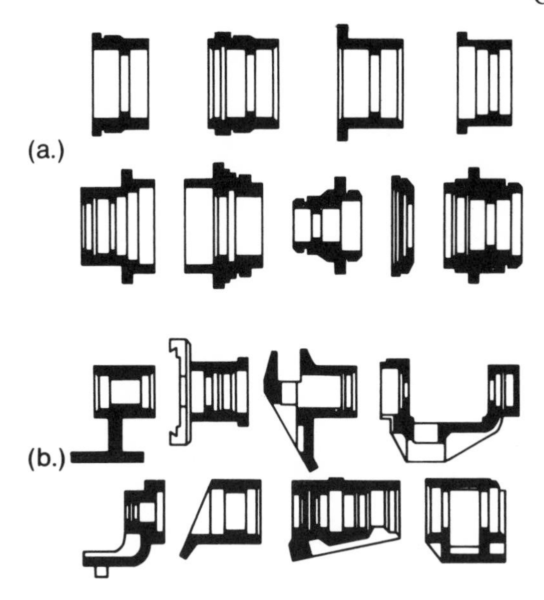

Figure 3.8. Group technology is a manufacturing philosophy that organizes similar parts into families such as (a) and (b) to standardize fabrication steps. (Courtesy of Westinghouse Electric Corp., Pittsburgh, Pennsylvania.)

steps to be standardized. Examples of part families are shown in Figure 3.8.

Most process planning systems now use retrieval techniques. In this approach, process plans are developed for part families from existing data bases on standard tooling and fabrication processes. One of the first of these systems was the CAM-I Automated Process Planning (CAPP) system. A commercial system called MIPLAN is also available from Organization for Industrial Research Inc.

Retrieval systems are far from the future generative process planning systems many experts predict. Generative process planning systems currently under development by CAM-I would produce process plans directly from the geometric model data base with almost no human assistance. The process planner would review the input from design engineering on a CAD/CAM terminal and then enter this data into the computer system, which would automatically generate complete plans.

ROBOTICS

Robots are automated manipulator arms that perform a variety of material-handling tasks in the CAD/CAM system. Robots may select and position tools and workpieces for NC machine tools. Or they may carry equipment or parts between various locations on the shop floor. They also may use their mechanical hands (called end effectors) to grasp and operate drills, welders, and other tools.

Most robots presently are programmed in a so-called teach mode. In this approach, a user physically leads the robot through the individual steps of an operation. This type of manual teach-programming is time-consuming and error-prone. Also, program changes usually require the entire sequence of steps to be repeated.

To overcome these difficulties and extend robot capabilities beyond the traditional teach mode, present research in robotics is directed at developing advanced programming languages with which robot instructions may be issued through a computer. One of these advanced languages presently under development is the IBM Automated Parts Assembly System (AUTOPASS). This language attempts to eliminate the need for issuing detailed instructions to the robot. The program automatically determines the grip points and motion paths from the geometric data base.

Some languages are intended to operate with artificial sensory input that enables the robot to act more independently. For example, the Stanford Research Institute Robot Programming Language (RPL) includes capabilities for interpreting video signals, enabling the robot to visually identify parts. And Draper Industrial Assembly Language (DIAL) developed at Charles Stark Draper Laboratory uses electronic force-feedback to duplicate human sense of touch in assembling components.

Advanced robot systems are also being developed as part of the U.S. Air Force ICAM project, where the goal is to organize all manufacturing steps around computer automation. As part of this program, a robot system has been developed to fabricate sheet-metal parts for the F-16 aircraft. The robot drills sets of holes to 0.005-in. tolerances with bits it selects from a tool rack and machines the perimeters of 250 types of parts. The robot system reportedly is four times faster than conventional manual fabrication.

Present plans for future CAD/CAM development have coordinated teams of robots and NC machine tools divided into group-technology work cells. The simple work cell in Figure 3.9 shows a robot tending two NC turning machines, performing much the same as a human technician.

Figure 3.9. Cincinnati Milacron T^3 industrial robot tends two NC-turning machines in much the same manner as a human operator. In this work cell, the robot places materials in the appropriate machine, removes and inspects finished parts at a laser gaging station, and then stacks them in piles for further routing. (Courtesy of Cincinnati Milacron Inc., Cincinnati, Ohio.)

FACTORY MANAGEMENT

Factory management ties together the other CAM areas to coordinate operations of the entire manufacturing facility. Factory management systems rely heavily on group technology, with individual manufacturing cells fabricating families of similar parts. And computers perform various management tasks such as inventory control and scheduling in material requirements planning (MRP) systems.

Predictions are that individual manufacturing cells ultimately will be linked together and controlled by a unified computer system, paving the way for overall factory automation. Production technology forecasts by some experts indicate that factories totally automated by computer will be a reality before the end of this century.

Total automation, however, does not mean a factory without people but rather one automated to the fullest extent practical. To implement automation efficiently, the CAM-I Advanced Factory Management System depicted in Figure 3.10 envisions a real-time distributed information system oriented on various levels of management.

The system is envisioned as being hierarchical in nature. At the highest level, an overall system manages the complete production cycle from raw materials to finished goods by controlling the activities of the many job shops producing parts for the final product. In turn, lower-level control centers manage the combinations of operators and machines that make up the individual work centers. In this manner, the system interfaces manual operations as well as highly automated robot and NC workstations.

In this heirarchical arrangement, each control center responds to the requirements received from the next higher level by issuing more detailed requirements downward and summarized feedback for reporting upward. This system is intended to provide each level of factory management with computer aides to plan events and allocate resources.

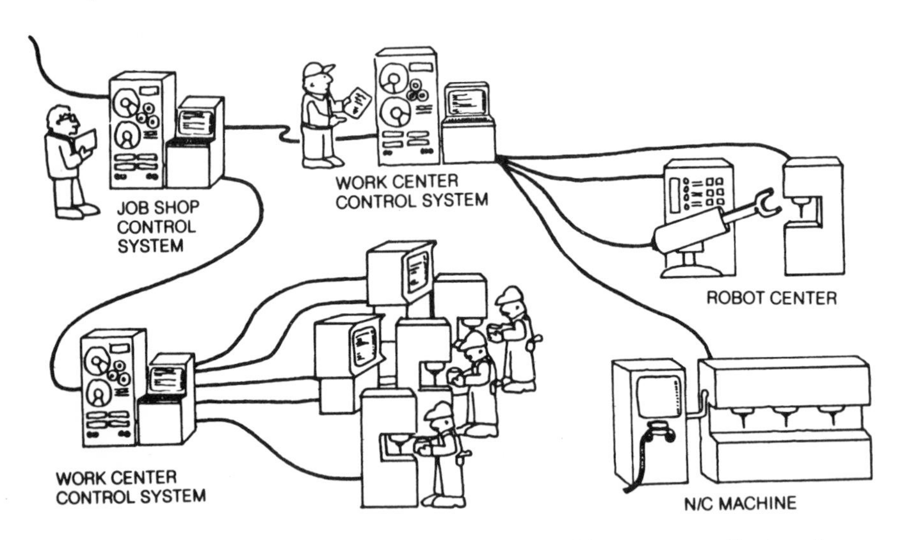

Figure 3.10. The CAM-I Advanced Factory Management System is a distributed network to provide each level of management with computer aides to plan events and allocate resources. (Courtesy of CAM-I Inc., Arlington, Texas.)

4

Geometric Models

Experts consider one of the most important features of CAD/CAM to be the geometric model—the representation of part size and shape in computer memory. The model is so critical to CAD/CAM because many design and manufacturing functions use it as a starting point. For example, the geometric model may be used as a framework to create a more detailed finite-element model of the structure. Or it may serve as an input for automated drafting to produce engineering drawings of the part. Furthermore, the geometric model may be used as a basis for generating NC instructions for making parts on automated machine tools or for producing process plans that outline the sequence of steps required to fabricate the complete assembly.

Geometric models typically are created with pictures drawn on the screen of a CAD/CAM graphics terminal. As a result, the user requires no knowledge of computers or programming to perform geometric modeling. Conceivably, models may be created without a graphics system. But this would be a tedious and time-consuming procedure compared with the ease and speed of entering and extracting computer data with a graphics system. As a result, the geometric pictures on CAD/CAM screens are now virtually synonymous with the geometric models they represent in computer memory.

The type of geometric model created depends on the capabilities of the CAD/CAM system and the requirements of the user. There are 2D models for representing flat parts, 2½D types for parts of constant section with no side-wall details, and full 3D models for representing the most general shapes. The 2D and 2½D models are adequate for many parts, but most research in geometric modeling is aimed at developing more sophisticated 3D modeling capabilities.

Most 3D modeling is presently done with so-called wire frames made of interconnected line elements. Wire frames (also called edge-vertex or stick-figure models) are generally the simplest models to create; so they expend relatively little computer time and memory. They also provide accurate information on the location of surface discontinuities on the part. But wire frames convey no information about the surface themselves, and they do not differentiate between the inside and outside of objects. As a result, a wire frame representation of a complex structure may be ambiguous and often leaves much interpretation to the user.

Many of these wire-frame ambiguities are overcome with surface models, which represent the next higher level of modeling sophistication. Surface models precisely define outside part geometries. As a result, they are useful in NC tape preparation and other tasks where definition of structure boundaries is critical.

Although features such as automatic hidden-line removal can easily make surface models appear to be solid, they generally represent only a shell or envelope of part geometry. This inability to represent the solid nature of the object may lead to difficulties in determining weight, volume, density, center of gravity, and other mass properties. Moreover, extensive user interpretation may be required to determine internal part details. For example, the same surface model may represent either a totally solid object or a thin-walled structure made of sheet metal.

The highest level of modeling sophistication is 3D solid modeling. In this approach, elementary cubes, spheres, and other so-called solid primitives are combined to create complex models. Another technique uses surfaces to characterize part boundaries and then applies additional routines to define the solid mass bounded by these surfaces.

The appearance of solid models may be similar to wire frames or partial surface models with hidden lines removed. But solid models are the only type that represents the true solid nature of the object in the computer. This facilitates computation of mass properties. And cross-sections may be cut through the model more readily to expose internal details.

Solid models represent the cutting-edge of technology in geometric modeling. Early systems were slow and relatively expensive. And they typically required extensive user expertise and computer-processing time. However, increasingly refined programs are overcoming these deficiencies, and solid modeling is expected to gain widespread use in CAD/CAM.

WIRE FRAMES

An operator at a CAD/CAM terminal creates a wire-frame model by specifying points and lines in space. The terminal screen typically is divided into various sections showing different views of the model. Or some systems use only a single view with a movable working plane. The terminal screen is used in much the same manner as a drawing board to create various views of the model. Unlike manual drafting, however, CAD/CAM provides automatic features to speed design.

Essentially, the lines comprising the wire frame are not manually drawn. Rather, the system constructs various line segments based on points specified by the user and commands chosen from a function menu. For example, the user may specify two endpoints and give the computer a LINE command to generate a straight line. Or a straight line may be generated by the computer parallel or perpendicular to another line, or tangent to a curve. Up to 40 such commands are available to produce straight lines on some systems.

The operator also may produce curved lines using similar techniques. Circles may be generated automatically from a center point and a radius, three points on the circumference, or tangent points on other curves. In this same manner, the user also may produce conics, which are complex curves such as ellipses, hyperbolas, and parabolas. Most CAD/CAM systems also can generate splines, which are smooth curves fit through a series of arbitrary user-specified points.

Many other features help the user create the model. For example, points and lines produced in one view may be projected automatically into other views. And specific details may be duplicated on the model at specified locations. Moreover, the user may temporarily erase selected lines from the screen without deleting them from computer memory. And erased lines can be recalled to the screen at any time. This capability is useful in working with complex models to clearly view the area under construction. Likewise, certain model areas may be enlarged to add minute details and later reduced to the proper size.

The step-by-step procedures for creating models vary according to system capabilities and the individual technique of the user. Most turnkey systems use a split-screen approach in which multiple views are displayed and manipulated simultaneously. And some systems use a single so-called trimetric with various working views called up one at a time as required. However, these approaches are merely different ways of graphically representing and manipulating the actual 3D model contained in computer memory.

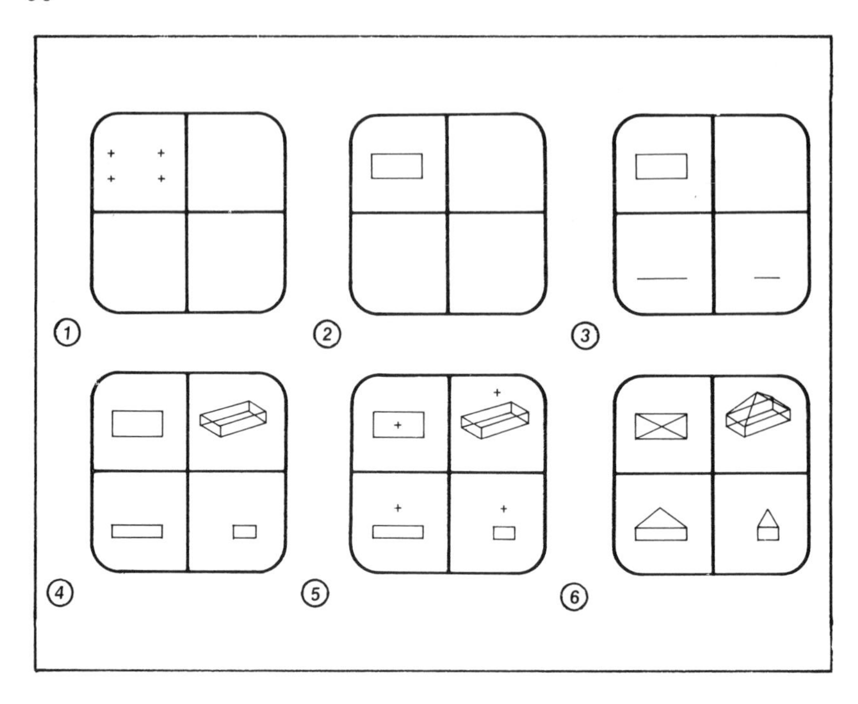

Figure 4.1. Sequence of displays from a graphics terminal shows how a simple wire-frame model is constructed using the split-screen approach. (Courtesy of *Machine Design Magazine,* Cleveland, Ohio.)

Figure 4.1 shows a typical sequence of steps to create a simple model with the split-screen approach. First, the screen is split into sections showing top, side, and isometric views. In this case, the user has specified four points on the top view representing vertices of that face. Second, the user commands the system to produce straight lines between the points, outlining the top face of the part. Third, the image is projected into the other three views. Fourth, the system projects the face into the second dimension to produce a rectangular block. Fifth, the user specifies a point above the block. Finally, the system connects lines between the point and the top corners of the block to create a pyramid and complete the model.

Complex networks of lines often make geometric models ambiguous and confusing to interpret; so some systems dash or delete hidden lines to make the solid geometry of the structure more apparent. Some systems require users to perform this task manually while others incorporate automatic hidden-line removal. For example, one system automatically establishes surfaces between line elements, and with a simple keyboard command the user either dashes or removes hidden lines as shown in Figure 4.2. This automatic feature presently is limited to flat planes, but hidden-line removal is under development for more complex surfaces such as cylinders.

Not all wire-frame modeling requires manipulation of points and lines

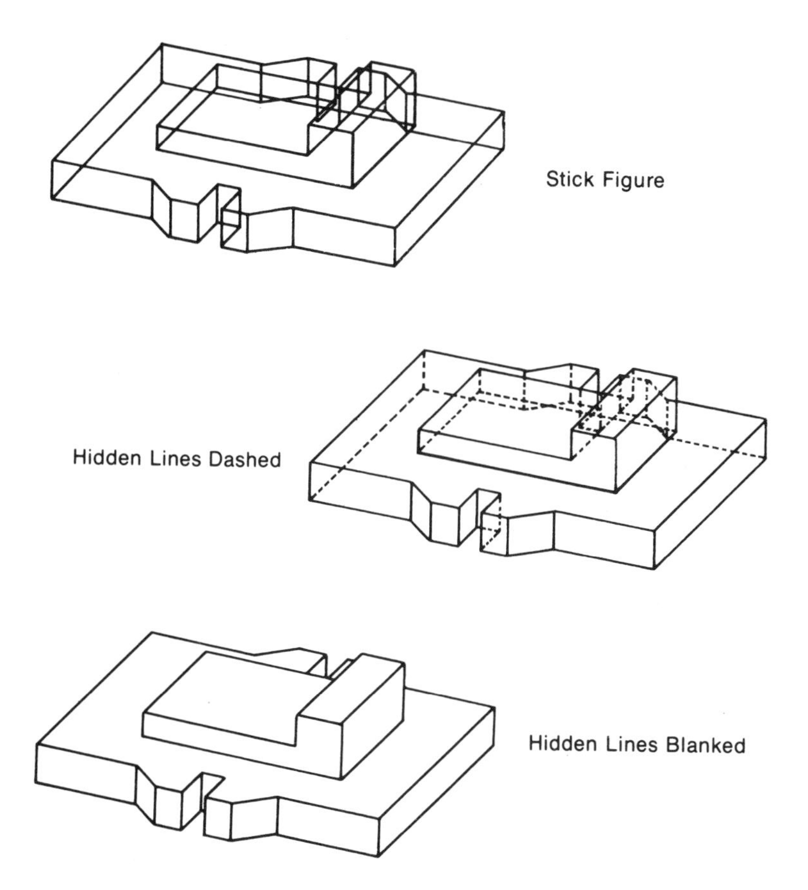

Figure 4.2. The solid geometry of a part modeled with a wire-frame representation is made more apparent by dashing or deleting lines that ordinarily would be hidden from view. (Courtesy of Computervision Corp., Bedford, Massachusetts.)

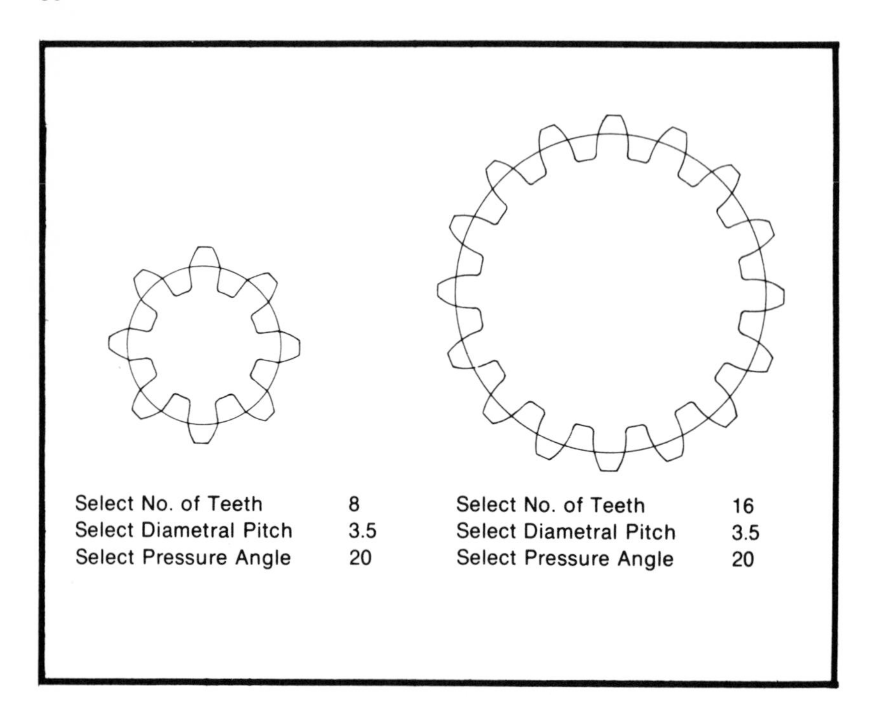

Figure 4.3. Gear models are created with Computervision's Parametric Element Processor from a few items of input data specified by the user. (Courtesy of Computervision Corp., Bedford, Massachusetts.)

on a CRT screen. Some systems automatically generate a geometric model from a few key parameters specified by the user. The system uses these values to scale and modify a base design of the part stored in computer memory. The system can design gears, as shown in Figure 4.3, as well as transformers, exhaust pipes, aircraft panels, and other parts where geometry depends on a given set of parameters.

SURFACE DESCRIPTIONS

The user creates a surface model by connecting various types of surface elements to specified line segments. The entire model may be made of these surface elements. But this approach is expensive and may provide more detail than necessary for many applications. As a result, some

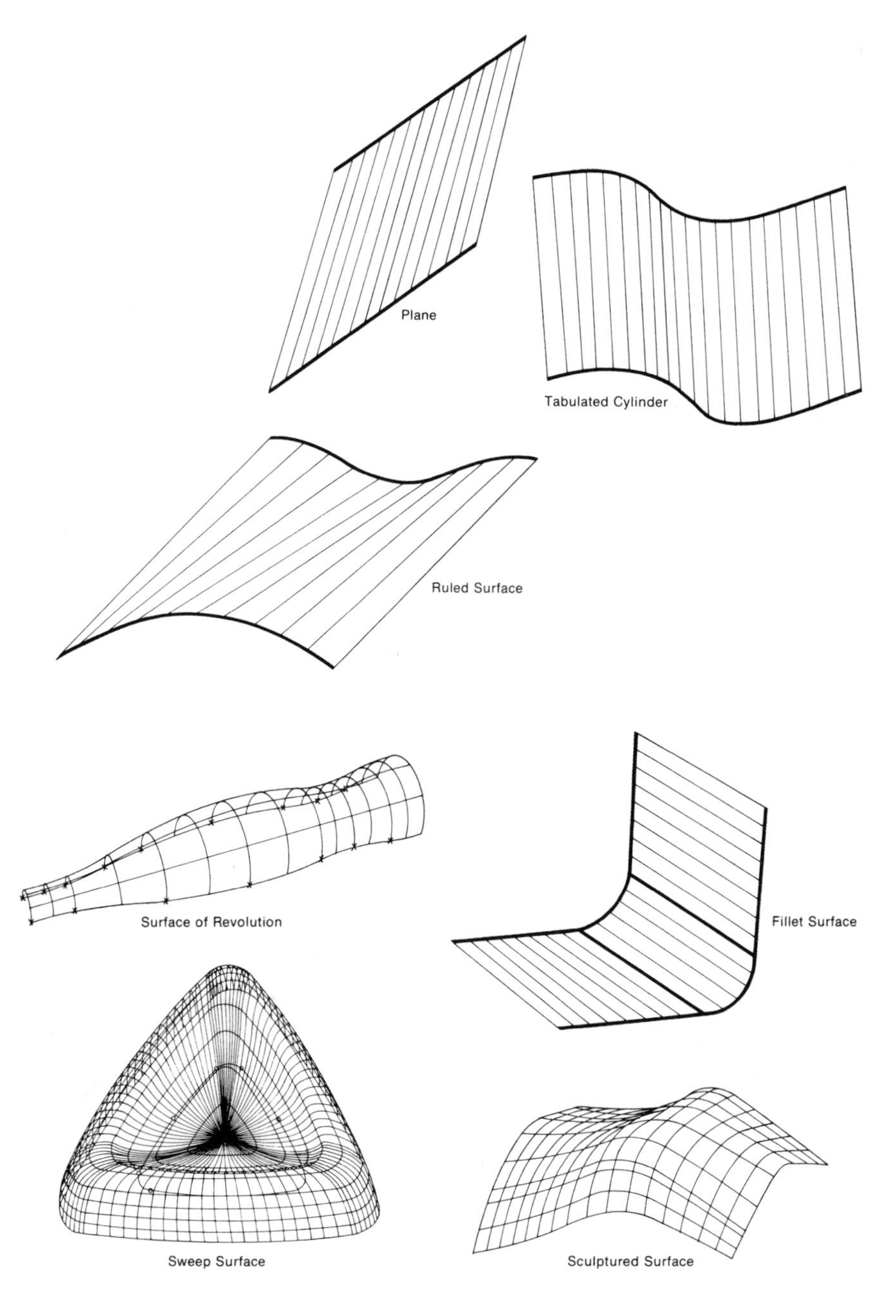

Figure 4.4. Surface types available for geometric modeling range from simple planes to complex sculptured surfaces. The surfaces are represented graphically as ruled lines or patch networks, but they are recognized as continuous surfaces in the computer. (Courtesy of CAM-I Inc., Arlington, Texas.)

models have surfaces for detailed faces and wire frames representing the rest of the part.

CAD/CAM systems typically provide extensive surface menus, which may include planes, tabulated cylinders, ruled surfaces, and surfaces of revolution, along with sweep, fillet, and sculptured surfaces. These surface types are presented in Figure 4.4.

Of course, the most basic surface type is the simple plane, which is merely a flat surface created by the system between two user-specified straight lines. A tabulated cylinder is made by projecting a free-form curve into the third dimension. This basically creates a curved plane between two parallel lines.

A ruled surface is generated between two different edge curves. Essentially, a surface is generated by moving a straight line through space with the endpoints resting on the two edge curves.

Revolving an arbitrary curve through a circle creates a surface of revolution, which is especially useful in modelng turned parts and other components with axial symmetry. An extension of the surface of revolution is the sweep surface, which moves an arbitrary curve through another arbitrary curve instead of a circle.

The fillet surface connects two other surfaces in a smooth transition. Creation of fillet surfaces formerly was a tedious and subjective manual operation. CAD/CAM systems now perform this task automatically with blending functions that provide consistent mathematical continuity between connected surfaces. These blending functions were first developed for designing aircraft and automobiles. But they are now used in CAD/CAM for virtually any structure requiring a smooth contour.

The most general representation is the sculptured surface, which is also called curve-mesh, free-form, cubic-patch, and B-surface. Sculptured surfaces are complex general contours such as an automobile body or helicopter blade that cannot be described with the usual lines and curves of conventional modeling.

Sculptured surfaces are generally represented by two families of curves that intersect one another in criss-cross fashion, creating a network of interconnected patches. These curve families do not have to be orthogonal, and the curve types are not fixed. In fact, individual curves in each family do not have to be parallel.

SOLID MODELING

In a typical solid modeling approach, the user produces a model by sizing, adding, and subtracting basic geometric solids called primitives. These are shapes such as spheres, circular and elliptical cylinders and cones, ellipsoids, rectangular parallelepipeds, wedges, and toruses. The approach assumes that most complex objects can be divided into these simple geometric figures.

One of the most widely used 3D solid modeling systems is SYNTHAVISION, a highly refined program containing over 12 primitives as shown in Figure 4.5. Initially, the program was prohibitively expensive and operated only on large computer systems. Consequently, use of the program mostly has been restricted to large corporations. But the program recently has been refined and is available on a more widespread basis, so use is expected to increase.

Inexpensive prototype solid modeling programs also are available from universities and research institutions. But these programs generally have fewer primitives. In addition, they are not as refined and have less software support than the commercial programs. There are probably more than 15 such programs developed around the world. The PADL program was developed at the University of Rochester under a National Science Foundation grant. And COMPAC was developed at the Technical University of Berlin. PROREN is a program from the Ruhr University

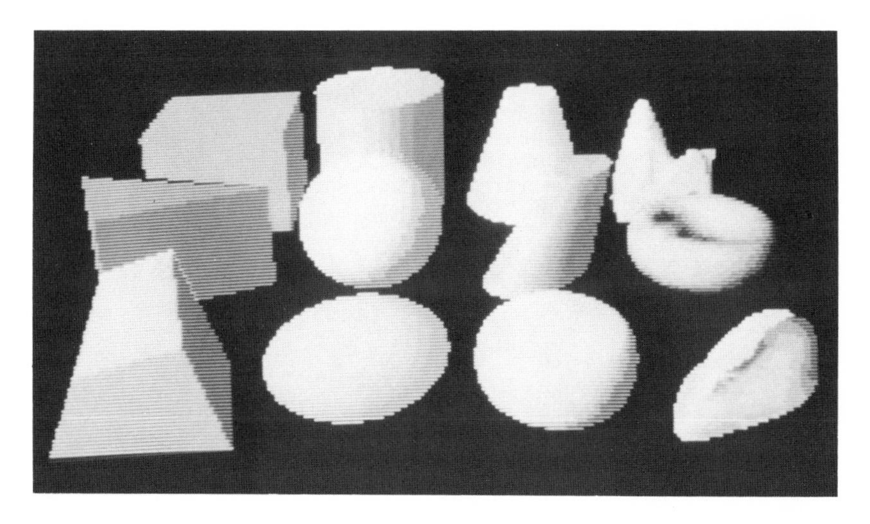

Figure 4.5. Menu of SYNTHAVISION primitives includes most shapes needed to create elaborate 3D models. With keyboard commands, the user selects the primitive, sizes and orients it on the screen, and adds or subtracts it from other shapes to produce a complete model. (Courtesy of Mathematical Applications Group Inc., Elmsford, New York.)

Brochum in Germany. And ROMULUS is from Cambridge University in the United Kingdom.

Another solid modeling approach uses an advanced form of surface modeling to fully define the solid nature of the object. In this approach, surfaces characterize the solid boundaries of the part, and then additional routines define the solid mass bounded by these surfaces. This approach is used in the PATRAN program developed by PDA Engineering and available from Digitial Equipment Corp. The program not only enables the user to create 3D geometric models, but also aids in the construction of complex finite-element models. The program also has extensive post-processing features for clearly displaying analysis results in graphical form. More than 12 of these solid modeling programs are now available, and use of the technology is expected to skyrocket in the next few years.

5

The Power of Finite Elements

CAD/CAM experts agree that one of the most powerful methods for analyzing a structure on a computer is probably the finite-element method. The technique determines characteristics such as deflections and stresses in a structure otherwise too complex for rigorous mathematical analysis.

In finite-element analysis, the structure is represented by a network of simple elements such as rods, shells, or cubes (depending on the geometry of the structure). Each of these tiny chunks of the structure has stress and deflection characteristics easily determined by classical theory. Thus, solving the resulting set of simultaneous equations for all the elements determines the behavior of the entire structure. Elements are connected at points called *nodes* that form a network known as a *mesh* or *grid.* The total pattern of elements representing the entire structure is called the *model,* one of which is shown in Figure 5.1.

The finite-element model typically requires hundreds or thousands of simultaneous equations, necessitating the use of a large computer and specialized data-handling programs. Furthermore, computer-assisted modeling and display aids usually are needed to build the model and interpret the results at a reasonable time and cost. As a result, a CAD/CAM system is virtually indispensible for finite-element analysis. The technique is too tedious and time-consuming to be performed manually.

One significant computer aid used extensively with finite-element analysis is graphics post-processing. In this operation, analysis data from the finite-element program is fed back into the computer, which condenses the multitude of information into visual form for quick interpretation.

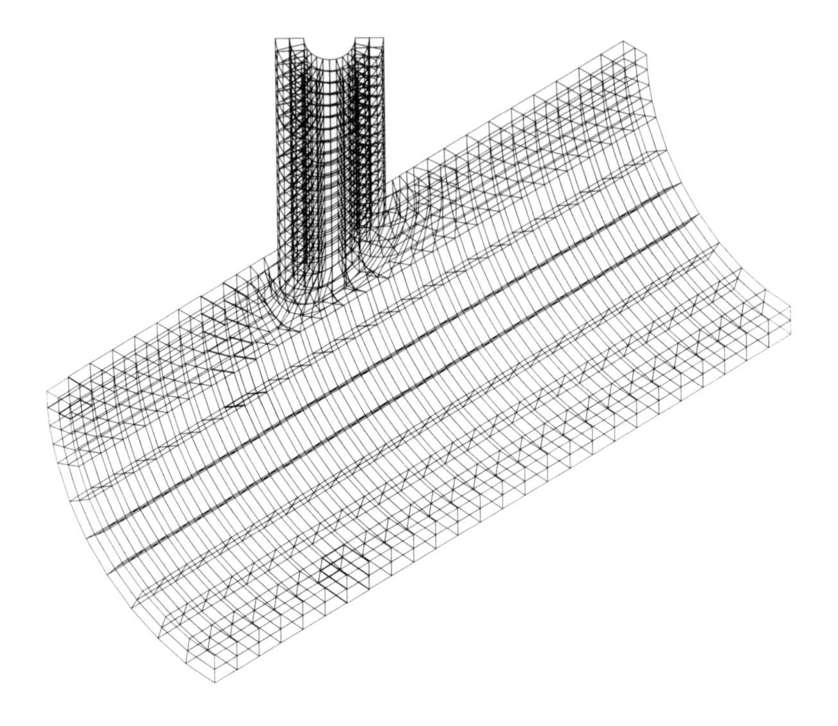

Figure 5.1. This typical finite-element model of a part was produced with ANSYS, a finite-element program for analyzing structures, heat transfer, thermal-fluid flow, thermoelectricity, and wave motion. (Courtesy of Digital Equipment Corp., Marlboro, Massachusetts.)

This capability is essential to convey the enormous volume of data that would be virtually indecipherable in tabular form.

The usefulness of graphics post-processing is illustrated with the analysis of a diesel piston cross-section shown in Figure 5.2. Contour lines show constant stress values displayed on a plane sliced through the component. Alternatively, stress levels may be indicated with color plots superimposed over the part outline.

Until recently, the finite-element method was so expensive that it was restricted to industries such as aerospace and nuclear that required precise analysis regardless of cost. But now, mainframe computers and special programs required to do the analysis are being offered at low cost through timesharing and leasing. Minicomputers and terminals are more powerful and less expensive than ever. And modeling techniques have been developed to provide accurate results with lower labor and computer-processing costs.

With these new features, finite-element analysis is spreading rapidly throughout general industry. Most new automobiles are analyzed with the finite-element method. And the technique is being applied to construction

Figure 5.2. Plot of diesel piston cross-section shows stress contours derived and displayed by graphics post-processing. (Courtesy of Structural Dynamics Research Corp., Milford, Ohio.)

machinery, agricultural equipment, pumps, air compressors, machine tools, home appliances, electric motors, fans, turbines, and numerous other products. The finite-element method is even being applied to many solid components that were formerly considered too complex for rigorous stress analysis. These include parts such as engine blocks, engine heads, and manifolds.

TEST PLUS ANALYSIS

One of the most important aspects of finite-element analysis concerns the combination of experimental and analytical data into a single computer model. For example, elastic components like tires, shock absorbers, and isolation mounts can be characterized more readily by testing them in a laboratory than by trying to make mathematical models of their dynamic properties. On the other hand, automotive frames, body sheet metal, and other rigid parts can be characterized more readily through analysis than through testing. The challenge has been in combining these two types of characterizations so that the behavior of a total machine or vehicle can be predicted before a prototype is built.

After years of work, software experts finally were able to develop workable programs that bring together analytical and experimental data. As a result, the performance of complex mechanical systems now can be evaluated while they are still in the early stages of design. For example, riding qualities, handling characteristics, and body rigidity of automobiles can be assessed before a prototype is built.

Closely tied to finite-element analysis is a technique called modal analysis. In this method, experimental data is gathered from various points on a structure to determine how it vibrates and deforms during operation. Two motion transducers typically provide input signals to the modal analysis system: one attached to the test structure and the other to the excitation source. The exciter induces structural vibration with a single impulse delivered with an instrumented hammer or with a range of frequencies delivered with a shaker. The structure is excited while the analyzer measures its response at a number of points.

The system collects and correlates the data from all the points. So-called mode shapes are then displayed showing how the structure deforms at its principal natural frequencies. These deflections occur rapidly (typically 20 to 30 times per second) and are generally small compared to the overall size of the structure. Thus, the system produces a slow-motion animated display in which the twisting, bending, and rocking of the structure are greatly exaggerated for easier interpretation.

Complex structures such as automobiles and aircraft are routinely studied with modal analysis as shown in Figures 5.3 and 5.4. Mechanical systems like these normally experience heavy vibration, and modal analysis can indicate where structural members should be reinforced. Also, lighter structures such as home appliances may use modal analysis to determine resonant frequencies and minimize noise transmission.

In the most sophisticated analysis, data from finite-element analysis, modal testing, and other empirical techniques are combined into a so-called system model to accurately predict how the structure will behave during operation. These applications involve massive amounts of data organized in huge matrices in the computer. Manually manipulating this data would be a horrendous task, involving much time and prone to human errors. So computer programs developed specifically for such data handling are necessary. One program known for this type of matrix manipulation is NASTRAN, which was originally developed by NASA for handling the huge amounts of data associated with analysis of stress and other dynamic characteristics of space vehicles.

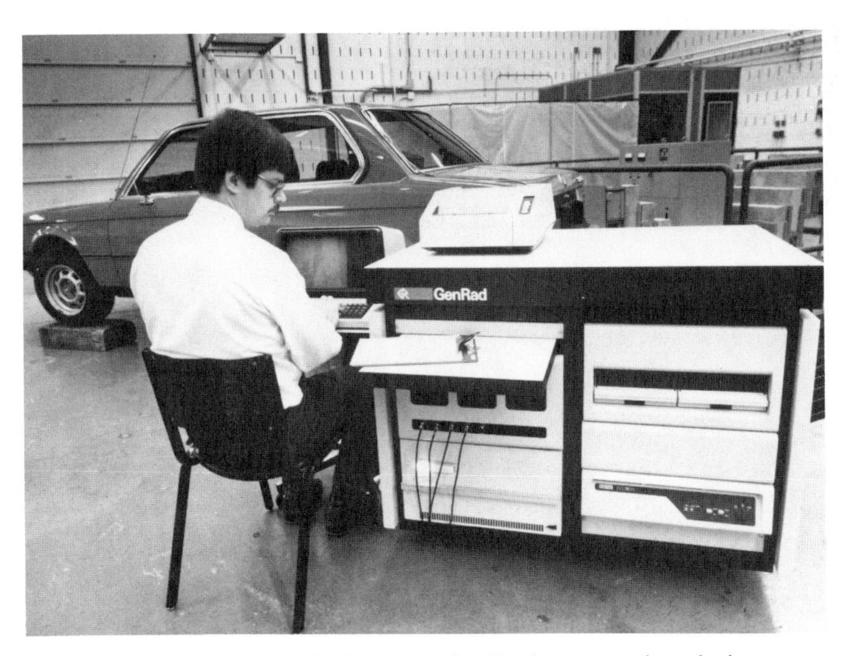

Figure 5.3. Analyst keys in data to a GenRad structural analysis system for modal testing of an automobile. (Courtesy of GenRad Inc., Santa Clara, California.)

The program accepts data in submatrices representing various components. These submatrices are then combined by the program to characterize the entire structure as a single system model. Additional load data is then fed in to simulate various forces the structure is expected to encounter during operation. For example, load data in a typical automotive application may be for turning, braking, or tire impact with a hole. By manipulating the matrices of data representing the structure and loads, the program is able to predict the response of the structure to these conditions.

An analysis of a pick-up truck is shown in Figure 5.5 to demonstrate the combination of experimental and analytical data into a computer model. In this application, vehicle components are divided for individual testing or analysis. The data is then combined into a system model, which predicts structural distortions for various operating frequencies and driver displacement resulting from a wheel unbalance.

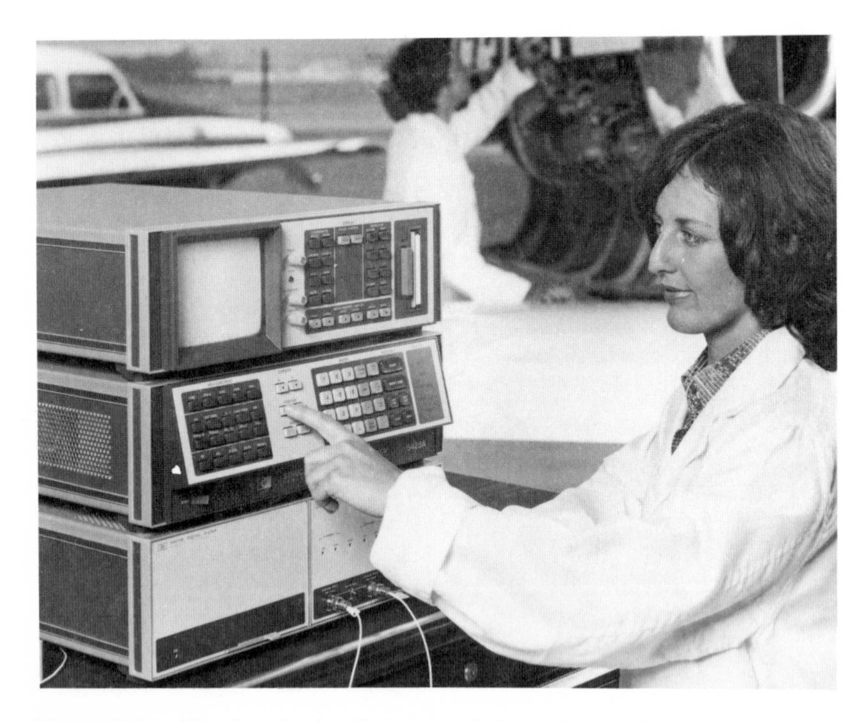

Figure 5.4. Hewlett-Packard structural dynamics analyzer is used in the modal analysis of a small jet aircraft. (Courtesy of Hewlett-Packard, Santa Clara, California.)

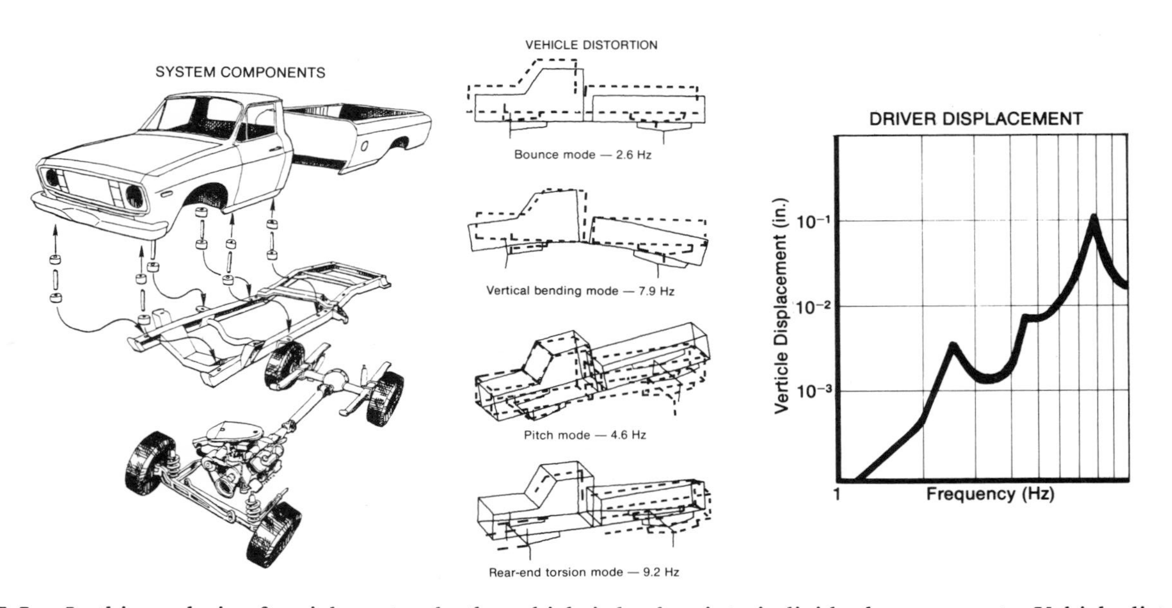

Figure 5.5. In this analysis of a pick-up truck, the vehicle is broken into individual components. Vehicle distortion and driver displacement are predicted by the computer model. (Courtesy of Structural Dynamics Research Corp., Milford, Ohio.)

BUILDING THE MODEL

Finite-element modeling experts recommend an iterative technique in which coarse models with few elements are increasingly refined with more elements in critical areas until the required accuracy is reached. This is probably the surest method of arriving at the simplest possible model that still yields satisfactory results.

In the analysis of a complex structure, a coarse model resembling a stick figure is typically constructed of simple beams. Rather than representing the true geometry of the structure, the beam model represents only how the hardware reacts to loads. Consequently, the beam model is used to determine overall deflections, to pinpoint probable areas of high stress, and to define general loading in the structure.

Beam models are constructed by first placing node points at each change in cross-section of the structure. Adjacent nodes are then connected with beam elements, each of which is assigned material properties (such as Young's modulus and Shear modulus) and stiffness properties (such as moment of inertia, torsional constant, and shear-area ratio). From all this data, the computer calculates the deflection of the structure at each node point. This information is then used to readily determine overall structural deformation and the resulting internal force distributions.

Computer costs to analyze a beam model typically are low enough (less than $20) to make these coarse models good for evaluating overall product design. Typically, several design options may be compared and the best one selected for further development.

After the overall structural characteristics have been determined from the beam model, refined fine-mesh models may be developed with more elements in critical areas with large deflections. This permits a more detailed analysis in those areas where stress and deflection need to be determined more precisely. The mesh density in these critical areas is increased until it reaches the required level of accuracy, and a coarse grid with fewer elements is used to represent the rest of the structure. In this way, expensive computer-processing time is not expended to analyze the structure in noncritical areas. Figure 5.6 shows a finite-element model of a block with a fine mesh of elements on a corner where high stress was anticipated.

Essentially, the aim of such an iterative modeling strategy is to minimize modeling time and computer costs while getting sufficient accuracy. And analysts use a variety of approaches to determine where the model demands a fine mesh and where coarse grids will suffice. For example, a graphic display can be used to pinpoint model areas that have high kinetic

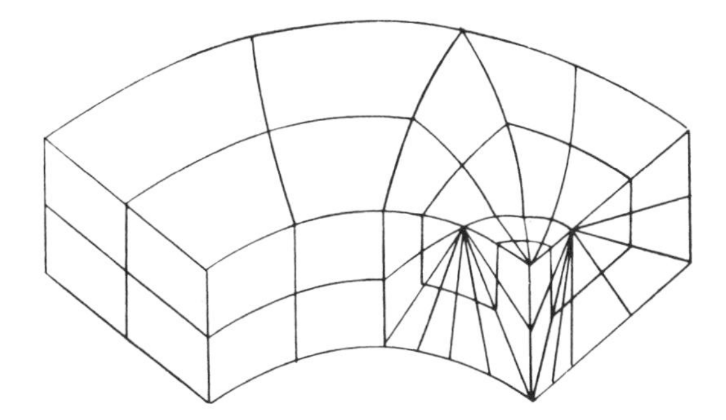

Figure 5.6. Finite-element model of a cracked block has a concentrated grid density in the corner where high distortion and stress are anticipated. The MENTAT software package that produced this model uses hexehedral elements that have straight or curved sides. (Courtesy of Marc Analysis Research Corp., Palo Alto, California.)

and strain-energy densities. Grid density in these areas then may be increased until strain energy is evenly distributed over most of the elements.

The effectiveness of this technique was studied at Structural Dynamics Research Corp. by means of a model of an automobile wheelwell. This was a thin-walled structure containing 39 holes, several ribs, and intricate changes in curvature. Wheelwell models and a graph comparing the number of elements with the modeling time and error are shown in Figure 5.7. Initially, the model was covered uniformly with 142 thin-shell linear and parabolic elements. However, strain plots indicated high energy concentrations in several areas, particularly around a hole in the structure's side. Elements were added to these areas for increased accuracy. But plots indicating some remaining high strain around the hole led to the addition of still more elements, producing a 520-element model. In this manner, adding elements only in the critical areas permitted an accurate representation of the structure with a minimum number of elements.

In comparing analytical predictions of wheelwell mode shapes and vibration frequencies with test results, analysts found the error to be 10% for the 520-element model. This was considered a modest error for a complex thin shell. Modeling time was over 160 hours for the 520-element model and increased linearly with the number of elements. Computer costs for the model were over $600. However, computer costs rise exponentially with the number of elements and would have been several

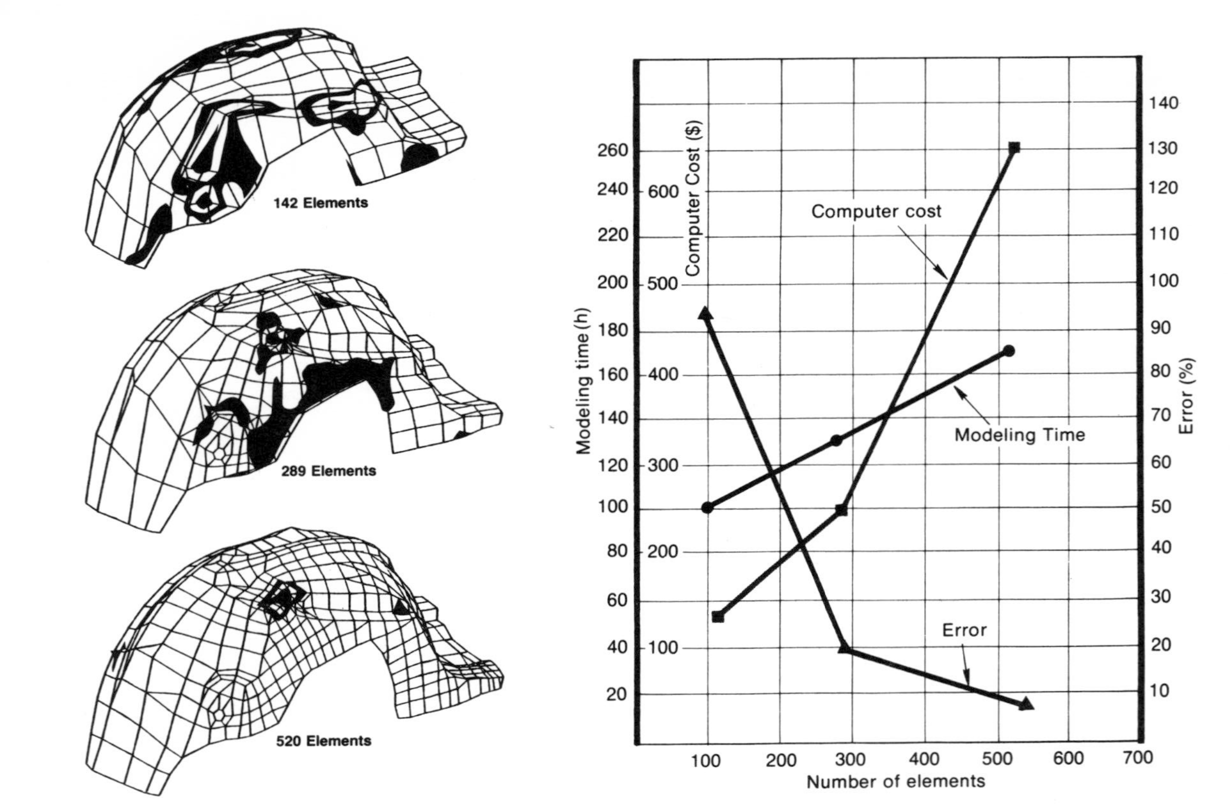

Figure 5.7. Analysis of a wheelwell was done with increasingly refined models. As the number of nodes in the model increases, error is reduced but modeling time and computer costs rise sharply. (Courtesy of Structural Dynamics Research Corp., Milford, Ohio.)

thousand dollars had a fine-mesh grid been applied over the entire model instead of only selected areas of high strain.

This iterative modeling approach is used extensively in the automotive industry to predict the operational performance of vehicle designs. The beam model simulates coarse behavior of the vehicle to a variety of road conditions. And the detailed models of components determines strain energy, stress concentrations, and deflections in individual components. A typical beam model of an automobile and fine-mesh models of various components are shown in Figure 5.8.

In the analysis of an automobile, components such as batteries are not studied individually in this manner but are represented as lumped masses. And the mass of nonstructural items like undercoating and trim is distributed over the entire model. For predicting overall structural responses, seats are depicted as rigid bodies and the passengers as lumped masses on springs representing seat-cushion stiffness. For predicting ride quality in a more detailed analysis, seats are modeled with finite elements to identify uncomfortable vibrations.

Automobile sheet-metal surfaces are often modeled with isoparametric elements. This accurately represents the curved surfaces with a minimum number of elements. Whereas conventional elements have nodes only at the corners, isoparametric types have additional nodes at the midsides. This enables the user to more accurately describe mechanical deformations and structural geometry, especially curved surfaces.

A variety of elements are available for modeling a range of surface types as shown in the Figure 5.9 library. Plane elements are used for flat, thin-walled areas. Shell elements generally represent thin curved surfaces. And thick-shell or brick-like solid elements are for areas with appreciable thickness.

In general, isoparametric models may require only about 25% as many elements as conventional models. This can reduce model-construction time, computer-memory requirements, and computer-processing time by as much as 50%. This sort of reduction is illustrated in Figure 5.10, where a truck wheel modeled with 713 conventional elements required 8 days to develop and consumed 1,531 seconds of computer-processing time and 150k of memory. The same part modeled with only 42 isoparametric elements required 3 days to develop and consumed only 483 seconds of computer-processing time and 70k of memory. Similarly, a diesel piston modeled with 722 conventional elements required 25 days to develop and consumed 5,609 seconds of computer-processing time and 320k bits of memory. The same part modeled with only 145 isoparametric elements required 16 days

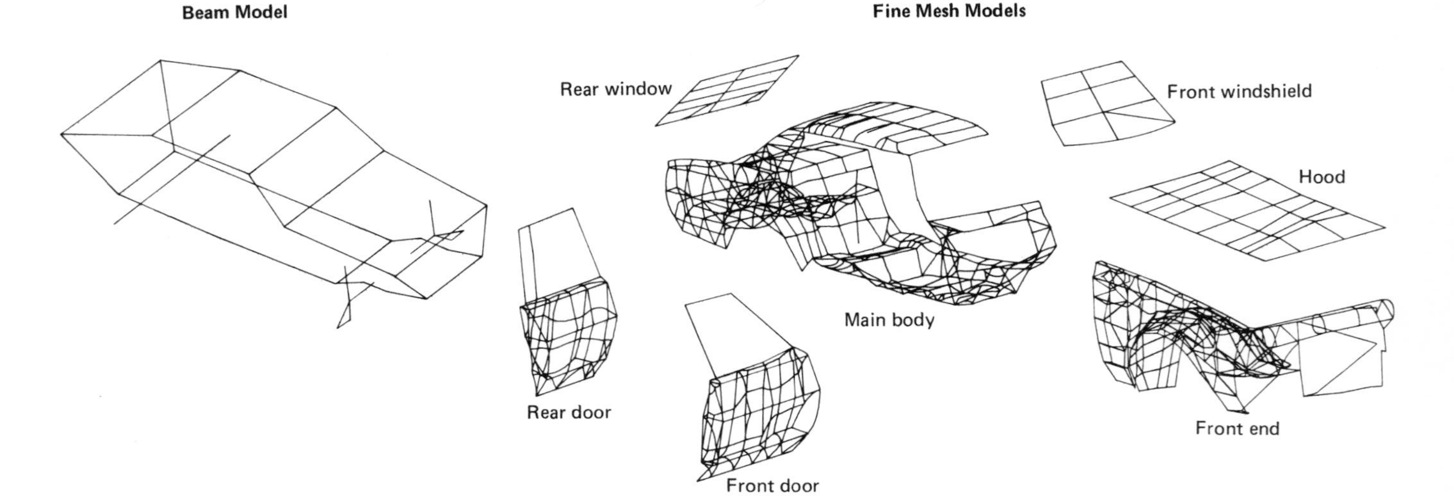

Figure 5.8. Beam model of automobile simulates coarse behavior of the vehicle to various road conditions. Detailed fine-mesh models predict stresses and distortions in individual components such as the body, hood, and doors. (Courtesy of Structural Dynamics Research Corp., Milford, Ohio.)

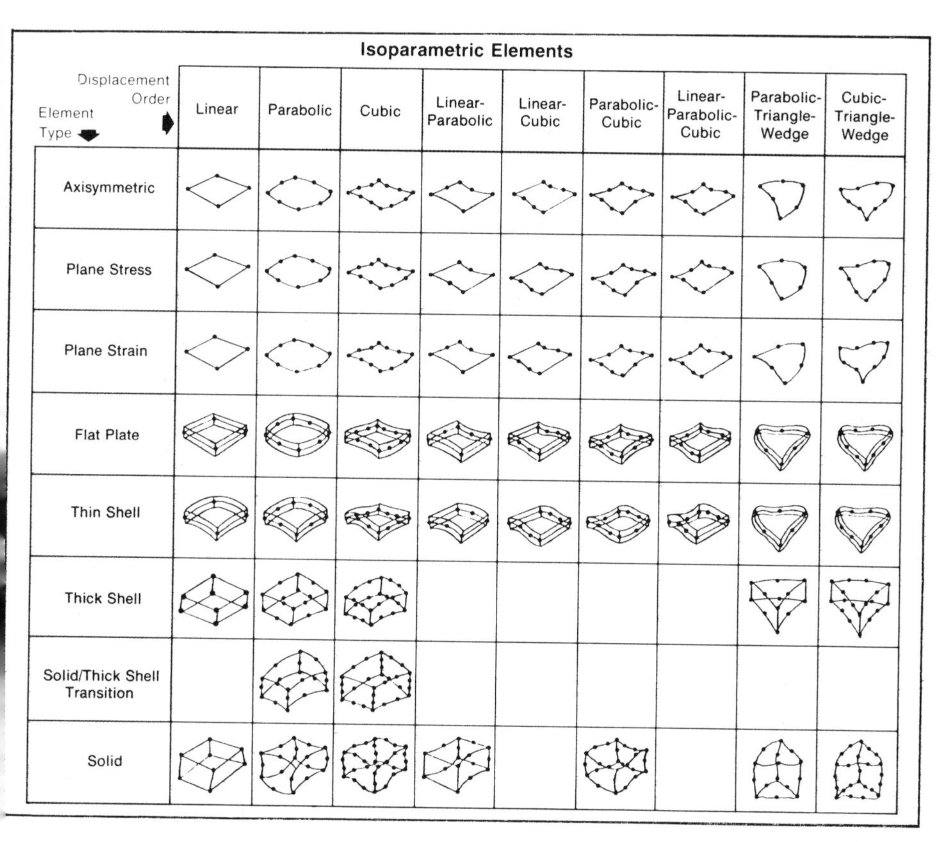

Figure 5.9. Library of isoparametric elements enable the user to more accurately model structural geometries with comparatively fewer elements, especially curved surfaces. (Courtesy of Structural Dynamics Research Corp., Milford, Ohio.)

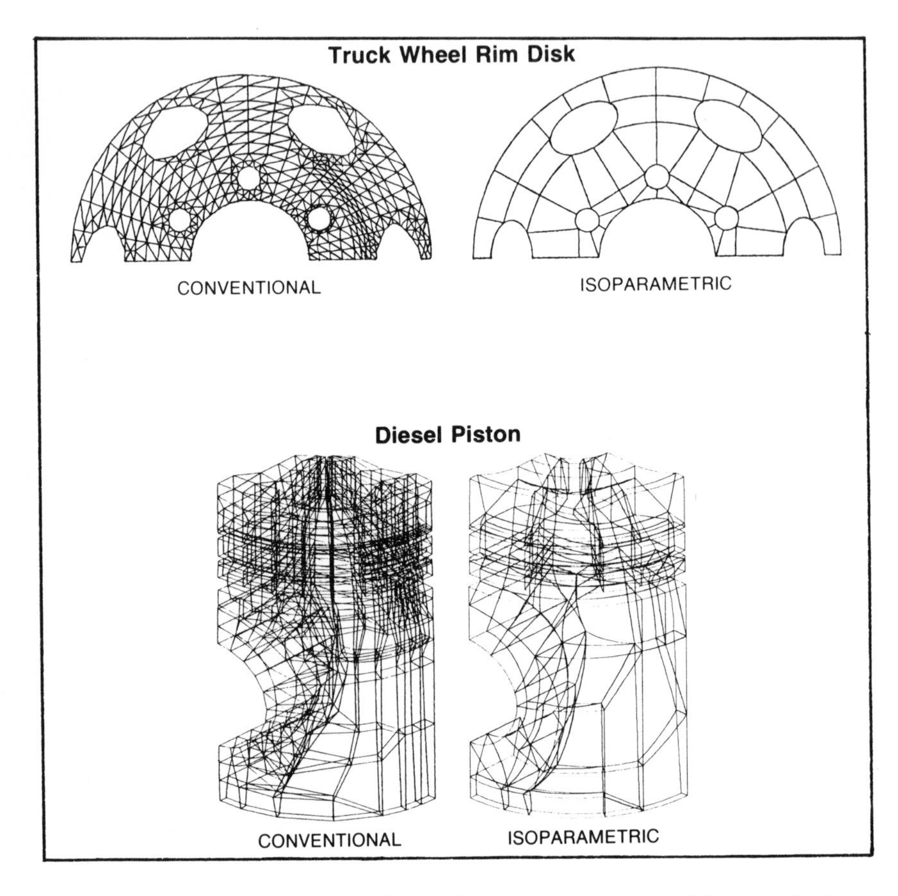

Figure 5.10. Comparison of wheel rim and piston models reveals that models constructed with isoparametric elements may require only about 25% as many elements as those made with conventional elements. (Courtesy of Structural Dynamics Research Corp., Milford, Ohio.)

to develop and consumed only 2,077 seconds of computer-processing time and 140k bits of memory.

Computer time and modeling effort may be reduced further by taking advantage of symmetrical characteristics of the structure. When a structure and its loadings and boundary conditions are symmetric, the resulting stress field usually reflects this same symmetry. Consequently, only one symmetric portion of the structure needs to be modeled, since no additional information is gained by reproducing identical stress fields across the planes of symmetry.

Some analysts invest savings from symmetrical modeling in an increased number of elements to yield greater accuracy. Taking advantage of symmetry in this manner often allows a two- to four-fold increase in the number of elements with no sacrifice in modeling cost or time. Furthermore, less computer-processing time and memory-storage space is required if symmetry is utilized.

Most finite-element programs have data-generation algorithms for handling repeated patterns of symmetric parts. Parts such as an automobile hood may have only two-side symmetry. But parts such as fans, flywheels, and other circular structures, comprised of many identical pie-shaped sections have even greater symmetry. Moreover, many parts have antisymmetric geometries which may be utilized. This is a more subtle form of symmetry in parts with "Z" shapes, for example.

MODELING AIDS

Nodal coordinates, connectivity, element definitions, and other model data are fed into the computer for finite-element analysis. The data may be manually written on lengthy data sheets and then transferred to computer cards by a keyboard. In the early days of finite-element analysis, this manual method was used exclusively because there was no better way. And today, data is still handled manually be some users—primarily beginners or those who have only a few simple parts to analyze. Manual data-entry requires relatively inexpensive equipment. But it is tedious, time-consuming, and error-prone. Errors are usually frequent in manual modeling because of the tedium in writing line after line of numerical data. As a result, most users rely heavily on computer-assisted modeling aids. Figure 5.11 shows an engineer using the UNISTRUC modeling program to produce a finite-element model of a rocket housing.

Computer-assisted modeling greatly increases modeling speed and accuracy. But errors still may be introduced when keys are incorrectly

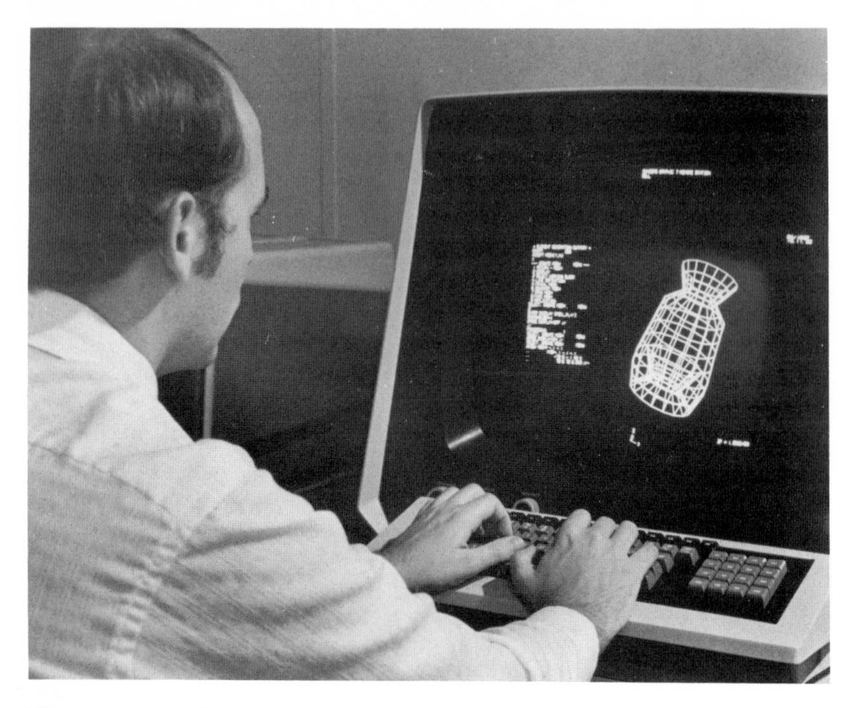

Figure 5.11. The UNISTRUC program allows engineers to create finite-element models of parts such as this rocket housing in less than an hour compared to ten days previously required using manual methods. (Courtesy of Control Data Corp., Minneapolis, Minnesota.)

punched or cursors improperly positioned. Most of these errors can be detected on the graphics screen that displays the model as it is being built.

The four most common errors are shown on a model of a support bracket in Figure 5.12. One element is missing as a result of a deleted line of connectivity data between two node points. The curved element was produced instead of the proper straight-line one because a mid-node coordinate was incorrectly entered. Gridwork is missing on the curved right edge of the part because the nodes and elements were not entered. And a normally rectangular set of elements is distorted because of a misnumbered node point.

Modeling aids include digitizing tablets and interactive computer graphics displays. This equipment permits an analyst to develop a model and feed the data into a computer in half the time required for manual modeling. Typically, a schematic drawing of a part is placed on a digitizing tablet and node points are entered into a minicomputer with a cursor. The minicomputer then constructs elements between these nodes according to operator instructions. In integrated CAD/CAM systems, the user may alternatively retrieve the geometric model from the data base and have the computer system produce the finite-element model with automatic mesh-generation routines.

After the finite-element model is built, the data is transferred to a remote mainframe computer for analysis. Depending on the type of analysis performed, the mainframe may provide stress, deflection, vibration modes, or temperature distribution. This output data is sometimes provided as numerical tabulations. But this bulk of numerical data is virtually unintelligible to anyone but an experienced analyst. More often,

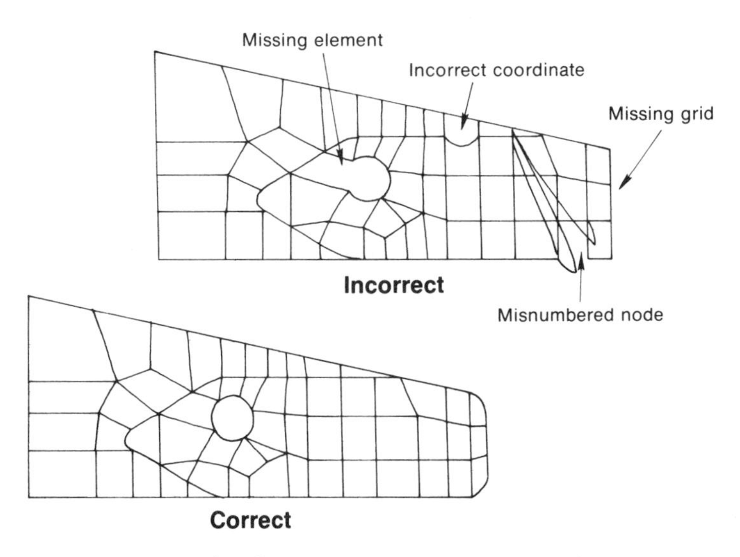

Figure 5.12. The four most common errors in generating a finite-element model are shown here. Errors are frequent in manual modeling, but may also happen with computer-assisted modeling because of incorrectly punched keys or improperly positioned digitizing cursors. However, input errors are easily detected in the visual display. (Courtesy of Structural Dynamics Research Corp., Milford, Ohio.)

computer graphics is used to produce pictorial outputs such as animated beam models, stress contour plots, or color-coded diagrams.

Today's sophisticated finite-element programs accept a variety of input data and present the computer output in a readily understandable format. As a result, they markedly lower modeling time and cost. For example, the SUPERTAB program demonstrated in Figure 5.13 reportedly reduces modeling time and cost by 75%.

Automatic digitizing and interactive graphics equipment required for computer-assisted modeling, however, are fairly expensive. Consequently, the high cost of this hardware must be weighed against the engineering time saved. Generally, equipment price rises with its sophistication. For a minimal initial investment, the new user may buy or lease a basic digitizing

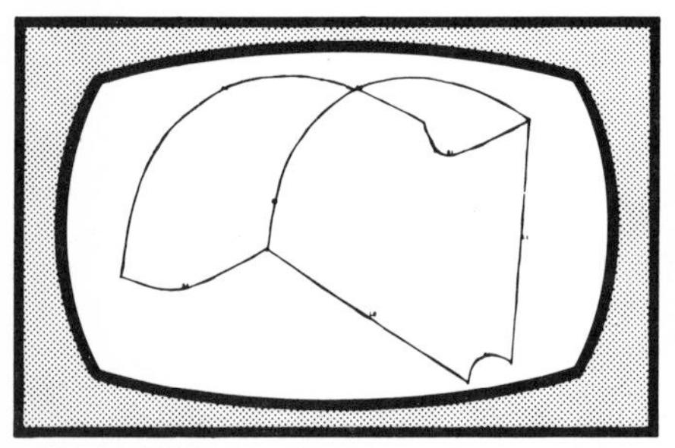

1. With SUPERTAB, the model geometry of a wheel is defined using points, lines, arcs, and splines.

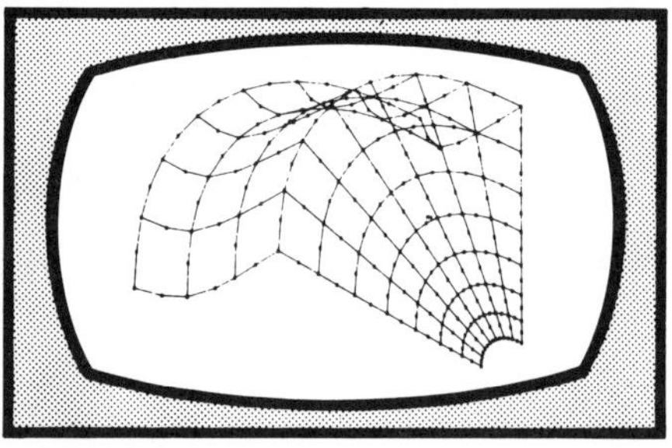

2. Automatic mesh generation produces nodes and elements within the same boundaries.

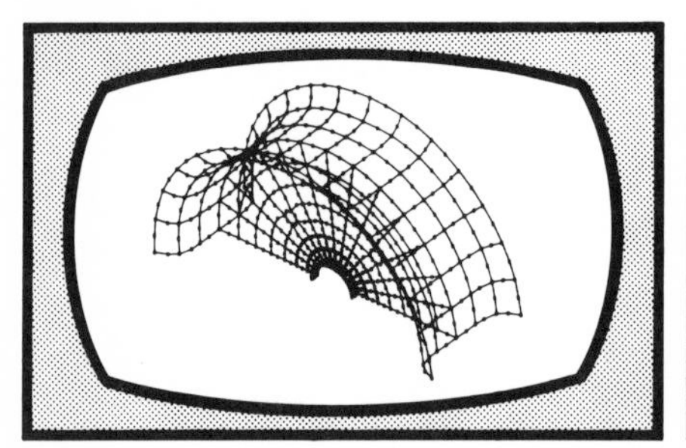

3. Multiple reflections quickly generate the rest of the model.

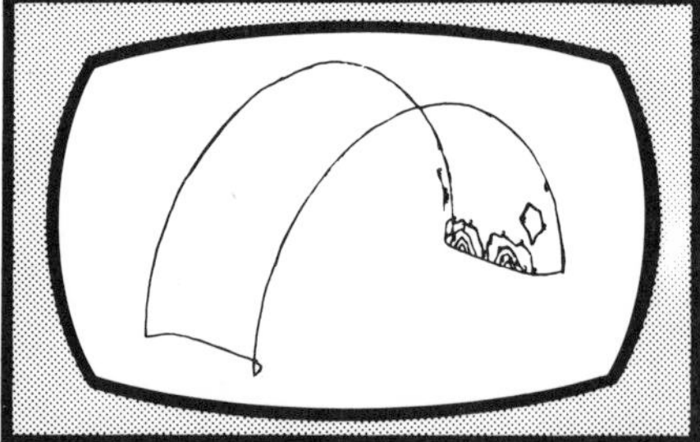

4. OUTPUT DISPLAY permits the visual interpretation of analysis results. Here we see the maximum principal stresses of the top surface.

Figure 5.13. Modeling aides such as SUPERTAB speed the modeling process and thus reduce costs. They accept a wide range of input data and present the computer output in a readily understood graphical format. (Courtesy of Structural Dynamics Research Corp., Milford, Ohio.)

tablet and graphics terminal connected to a time-shared host computer for under $20,000. The more experienced user may eventually justify the purchase of a complete minicomputer system for $75,000 to $100,000.

MORE THAN STRESS ANALYSIS

The finite-element method successfully tackles complex stress-analysis problems that were formerly impossible to solve. But the real beauty of the technique is that it is not confined to stress analysis or solid mechanics. Use of the finite-element method outside of solid mechanics began to appear about 10 years ago when researchers recognized the promise of the method in solving extremely complex problems in areas of continuum mechanics such as fluid mechanics. For example, analysts used finite-element models to evaluate fluid seepage through rigid and elastic porous media.

Hydrodynamic lubrication is one area of fluid mechanics where the finite-element method is now particularly useful. Analysts formerly had no accurate way to predict the behavior of lubricants under operating conditions. This deficiency was particularly troublesome in bearing design. The mechanical structure of the bearing could be analyzed precisely, but the effects of lubricants could only be roughly approximated. As a result, a bearing that appeared satisfactory in design sometimes performed erratically or failed prematurely in service.

There are specialized techniques for analyzing lubricant films on simple bearing surfaces under ideal conditions. But these methods are impractical for bearings with complex geometries such as grooves and recesses, especially when the oil film is subject to viscosity changes and other operating effects.

In contrast, the finite-element method deals easily with irregularities and variable boundary conditions. The same program can be used for any film problem and only the input data need be changed. Moreover, finite-element programs are fundamentally easy to write, and the procedure for deriving the finite-element matrix equations is fairly straightforward. In fact, computer programs for hydrodynamic lubrication closely follow existing programs for steady-state heat conduction, mass diffusion, and other phenomena governed by quasiharmonic equations.

The finite-element method is applied to lubrication theory in a fashion similar to that used for structural analysis. The film is represented by a network of small elements connected at nodes. Each node is assigned information about the film such as thickness, viscosity, or tangential surface motion. From this data, the finite-element program determines the film pressure and flow from classical theory (the Reynolds equation, developed in 1886).

The analyst then uses this information to determine lubricant operating parameters such as load capacity, stiffness coefficient, and damping coefficient. These lubricant parameters then may be coupled with the bearing characteristics to predict the behavior of the total bearing system. This may include factors such as vibration, frictional losses, surface wear rates, and service life. This means that engineers can evaluate a design or proposed revision quickly and inexpensively before the final bearing configuration is committed to hardware.

The finite-element method is used for virtually all types of bearings including grooved, hyperbolic bore, journal, step journal, spherical, squeeze-pad, and tilting-pad types. Generally, the bearings analyzed are in highly demanding applications with large loads and high speeds such as automobiles, engines, heavy off-highway equipment, and large rotating plant equipment.

Unlike finite-element programs for analyzing mechanical stress, those for lubricant films usually are not available off-the-shelf. Consequently, large companies such as General Motors which use the method extensively may develop in-house programs. But most users have neither the time nor the expertise to develop their own programs; so they generally obtain them from universities prominent in the field such as Cornell or the University of Virginia. A company may contract with an individual professor to develop a program for a particular problem. Or a group of companies may jointly support finite-element research at the university in exchange for first use of the programs that are developed.

A finite-element model of a lubricant film is built by piecing together elements in building-block fashion. Most film models are made of 2D triangular and quadrilateral elements. Simple 3-node triangles are used most often, but 6-node triangles and 8-node quadrilaterals sometimes are used for higher accuracy. And 3D tetrahedrons or hexahedrons often are used in more complex analyses.

A 2D model of the fluid film sometimes is combined with a 3D model of the bearing surface to represent a complex lubrication system. An example of this combination is the analysis of squeeze films between porous surfaces. The fluid film is represented by plane triangular elements and the porous bearing material by solid tetrahedral elements. These two element types account for standard lubrication effects as well as for the flow of lubricant through the porous region. The method is useful for analyzing bearings made of powdered metal, ceramic, and other porous materials. In addition it has been used to analyze the behavior of bone and cartilage in biomechanics.

Accuracy of the lubricant-film model depends predominantly on element position and grid density. One modeling study for a square bearing squeeze pad showed that analysis accuracy is highest when element diagonals are aligned with the direction of the pressure gradient. Then the elements more nearly represent the true pressure distribution throughout the film. And as with finite-element models for mechanical analysis, using fine-mesh grids only in areas where high accuracy is required appreciably reduces computer costs and modeling time.

Use of these techniques in the analysis of a typical lubricant film is shown in Figure 5.14. A typical pressure-dam or step-journal bearing has a rectangular step cut in the upper pad of the bearing to produce a stabilizing pressure rise and a relief track in the lower pad to accept excess oil. This bearing is easily analyzed for the finite-element method because complexities are accounted for with routine model features.

Each pad is divided into elements by an automatic mesh-generation routine. Grid density in the top pad is concentrated near the step where increased accuracy is required. Likewise, grid density for the bottom pad is greatest around the region of minimum film thickness. For maximum accuracy, element diagonals for both models are aligned with the pressure gradients direction.

The analysis determined the load created by the step and the dynamic coefficients of stiffness and damping. These parameters enabled analysts to predict the stability of the bearing during operation.

Most of the finite-element programs for lubricant films in use today are highly refined and are used routinely to solve hydrodynamic problems where film temperature is constant and bearing surfaces are rigid and impermeable. Researchers now are attempting to reach the same level of sophistication in thermohydrodynamics, elastohydrodynamics, and porohydrodynamics. These more complex analyses must include the solution of classic fluid equations as well as another complete set of equations for additional variables such as film temperature, bearing elasticity, and bearing porosity.

The theory and resulting equations for these additional variables are well established by themselves. However, huge amounts of expensive computer time and memory are required to solve the large number of interrelated equations. Current research in finite-element film analysis is aimed at developing programs to solve these complex fluid problems economically.

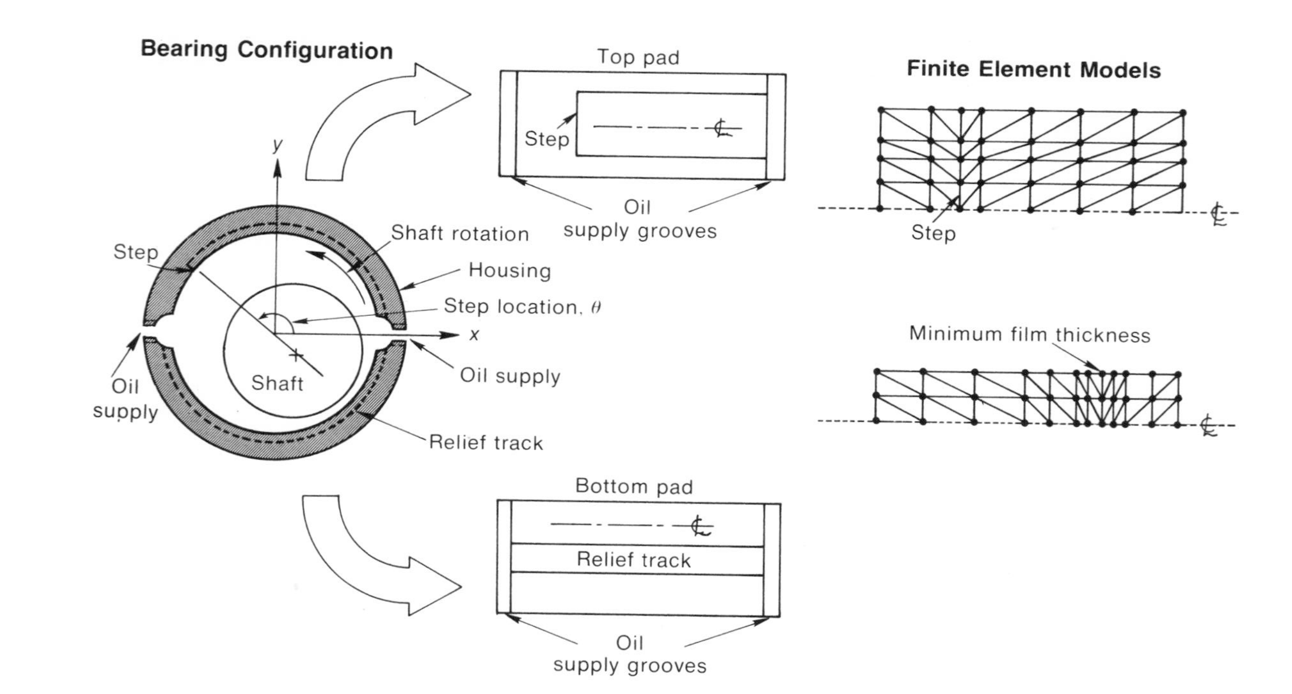

Figure 5.14. In this analysis of the lubricant film in a step-journal bearing, the grid density for the top pad is concentrated near the step where increased accuracy is required. Likewise, grid density for the bottom pad is concentrated around the region of minimum film thickness. (From Nicholas and Allaire, 1978, reprinted by permission of the American Society of Lubrication Engineers, Publishers of Lubrication Engineering and ASLE Transactions. All rights reserved.)

6
Manufacturing by Computer

The use of computers in manufacturing is growing rapidly. Some of these uses involve the transfer and interpretation of information for factory managers. These systems act as aids in running and maintaining the entire plant. Typical applications include inventory control, scheduling, machine monitoring, management information systems, and quality control. These applications represent important contributions of the computer for handling information and are gaining increasing importance in the management of manufacturing operations.

The actual control of the physical manufacturing process has equally great potential, but associated with it are difficult technical challenges. Much of the effort in this area is concentrated on developing more advanced computer systems for controlling NC machine tools. Most of this work applies to metal-cutting tools. But NC also controls flame cutters, arc and spot welders, fabric cutters for the shoe and garment industries, composite material lay-up equipment, wood-working machinery, and many more classes of tools.

Closely associated with these efforts in NC is the work in developing sculptured surfaces software to machine more generalized shapes. In addition, researchers continue to develop improved robotic systems. The most recent advances in this area are the artificial senses that give automated manipulator arms senses of vision and touch. Research in these areas is aimed ultimately at combining refined, highly sophisticated machine-tool controls and robotic systems into future automated factories.

MACHINE-TOOL CONTROLS

Automated machine tools in CAM systems perform the drilling, grinding, cutting, punching, milling, and other operations that shape raw material into finished parts. One such machine tool made by Cincinnati Milacron is shown in Figure 6.1. From its set of thirty tools, the machine can perform a variety of operations such as the contouring cut depicted in Figure 6.2.

These machines are controlled with prerecorded, coded information called NC instructions written in control languages such as APT, which is considered the standard of the industry. In the most basic systems, NC instructions are stored on punched paper tapes that are placed on electromechanical tape readers hardwired to the machine tool.

More advanced systems may use computer numerical control (CNC), a scheme in which the machine is connected to a dedicated minicomputer that stores the NC instructions in its memory. CNC offers several advantages over standard NC. With CNC, instructions may be stored and handled more efficiently with a data cartridge or floppy-disk memory.

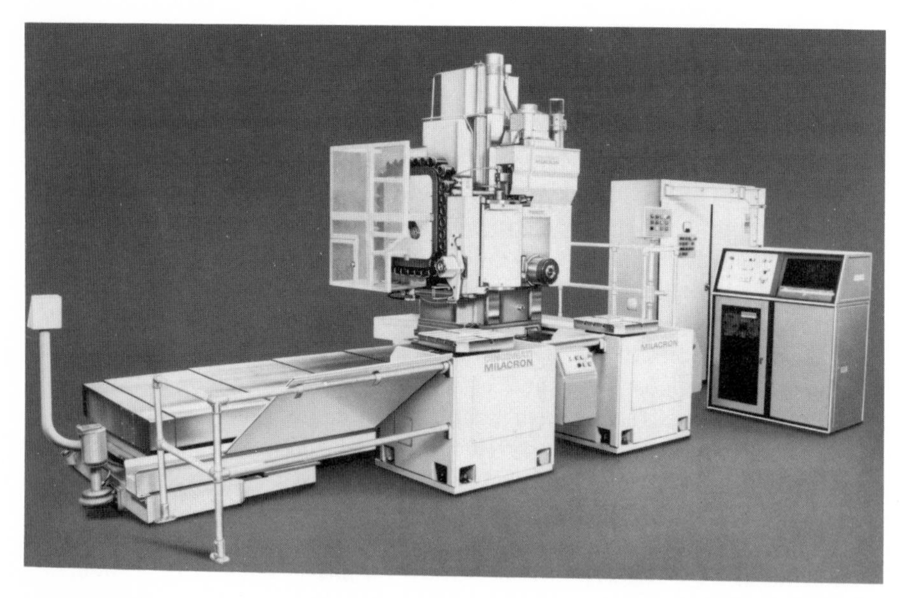

Figure 6.1. The CIM-Xchanger 20HC-2500 is a 20-hp machining center with automatic tool selection from a stored bank of 30 tools. (Courtesy of Cincinnati Milacron Inc., Cincinnati, Ohio.)

Figure 6.2. Machine tool in Figure 6.1 performs a contouring cut using a circle-diamond-square program to demonstrate machining capability. (Courtesy of Cincinnati Milacron Inc., Cincinnati, Ohio.)

Moreover, the CNC minicomputer system may be programmed to perform intelligent functions beyond simply controlling the machine. For example, machine-control data can be recorded and modified right at the CNC itself. Moreover, real-time and off-line diagnostic capabilities may be built into the CNC system. In addition, machining data and operator instructions may be displayed on a display screen on the CNC system.

The most sophisticated systems use direct numerical control (DNC). More recently, these systems have also been referred to as distributed numerical control, primarily because machine tool control is distributed among different computers. In this scheme, individual CNC units are linked to a central mainframe computer sometimes referred to as a supervisory computer. The mainframe supplies the part programs to the individual CNCs through communication lines, which also provide for feedback of production and machine-tool status from the shop floor. This allows for the coordination of machine-tool operation from a central location.

The DNC system permits rapid real-time feedback on problems as they occur so that corrective actions may be initiated as soon as possible. This feature is probably the greatest advantage of DNC systems. Another benefit of DNC systems is their capability to store massive amounts of program data and retrieve it rapidly. For this reason, DNC systems are used in aerospace and other large systems to store extremely long part programs. In addition, smaller companies may use the DNC system to store a variety of programs that may be used infrequently.

Several configurations of the machine tools and computers are possible in a DNC system. Some systems use a hierarchical arrangement of different levels of minicomputers performing various functions between the machine tools and mainframe. And still other DNC systems have eliminated the intervening minicomputers and have a direct interface between the central computer and machine tools. Some of these systems may have as many as 50 machine tools linked to a central mainframe.

When the mainframe is linked directly to the machine tools, two identical DNC computers sometimes are used. This permits the system to continue operating if one computer should fail. Rather than stand idle during normal system operation, the second mainframe can perform tasks such as batch computing, scheduling, or inventory that can be off-line for a time if necessary.

A typical DNC system developed by Allen-Bradley Co. is shown in Figure 6.3. The central computer in this system distributes part programs to the various machine-tool CNCs and manages a complex data network

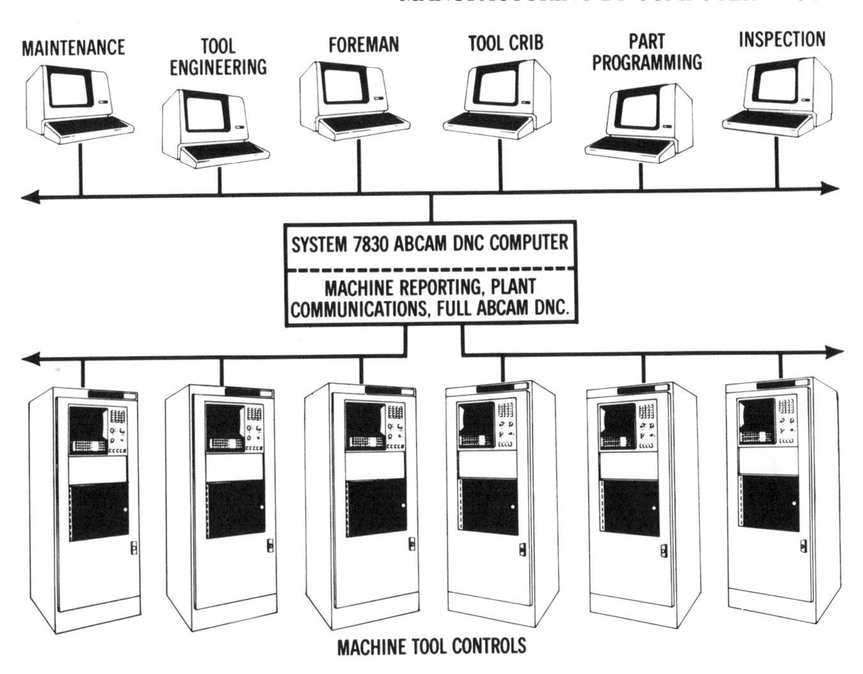

Figure 6.3. Allen-Bradley Series 7830 DNC System distributes part programs to multiple machine tools and coordinates a data network for communicating complex manufacturing information. (Courtesy of Allen-Bradley Co., Milwaukee, Wisconsin.)

that provides for machine monitoring and communication between various departments such as part programming, engineering, and inspection. By monitoring machine tools, the system stores status data and may generate reports on the number of parts per job, total time of job, set-up time, response time to correct malfunctions, downtime per machine, feedrates, and other data.

Future systems are being developed that do far more than execute standard NC instructions for making a part. This next generation of controls is expected to sense cutting conditions and adapt to them without intervention of a human operator. For example, such an adaptive-control system will be able to recognize a damaged tooth in a cutting tool in a few thousandths of a second. The system will stop the machine before the broken tooth comes around again, replace the broken tool, and restart the cutting operation—all without human supervision. Or the system may have capabilities for modifying an operation to compensate for temperature variations sensed by sets of thermistors in the machine tool or the workpiece. In addition, the system may make adjustments to the machine during cutting or grinding operations to correct for chatter and other vibrations sensed by accelerometers.

Perhaps the ultimate adaptive-control feature of these systems presently under development is the ability of the system to cope with unexpected problems. That is, the system may have the ability to determine corrective actions to unforeseen problems for which it was never programmed. The system will determine the appropriate corrective action by comparing the mathematical description (or signature) of the disturbance within the range of possibilities in its memory. This sort of machine intelligence and independent decision-making capability will be used extensively in automated factories of the future.

SCULPTURED SURFACES

The NC instructions that control machine tools in a CAM system typically are written by part programmers from engineering drawings or the geometric model provided by the CAD end of the system. These NC programs are not particularly difficult to write if the part contour consists of planes, cones, spheres, and other forms readily defined by mathematical equations.

In contrast, sculptured surfaces have arbitrary, nonanalytical contours that may not obey mathematical laws. For this reason, sculptured surfaces were formerly considered to be impractical to machine with NC. Rather, they have been defined traditionally by subjective curve-fitting techniques and made with time-consuming and expensive hand-finishing operations. Clearly, the sculptured surface is one of the most technically challenging aspects of CAM.

Contours that may be classified as sculptured surfaces are found in a wide range of products including aircraft, automobiles, construction and agricultural equipment, machine tools, appliances, cameras, instrument cases, and motor housings. Some companies (most notably in the automotive and aerospace industries) have independently developed NC sculptured surfaces capabilities. But many of these programs cannot guide a cutter reliably. And the software generally operates only on the most sophisticated in-house computer systems. In any case, these specialized programs are, for the most part, proprietary and jealously guarded by their developers. Of greater interest to general industry is the on-going cooperative work undertaken by a group of companies to develop a common program. This software is for generalized shapes and can be easily implemented on a wide range of computer systems.

The cooperative effort to develop a generalized sculptured surfaces capability is closely tied to the evolution of NC technology itself. In the late 1950s, a group of companies pooled their resources and developed APT (Automatic Programming Tool). This is the most sophisticated NC language still used today for programming machine tools. They called their cooperative association the APT Long Range Planning Program (ALRP) and sponsored development work to refine and improve the program. They assigned this task to the Illinois Institute of Technology Research Institute (IITRI).

Initially, the machining capabilities of APT consisted of straight lines, circles, and planes. Then in the early 1960s, these capabilities were extended to more complex shapes such as conic and quadric surfaces. But the ALRP was concerned with the inability of the program to express free-flowing shapes; so in 1968 they sponsored a Sculptured Surfaces Project at IITRI. Their goal was to develop an APT-compatible program that could define these arbitrary shapes and make them with NC machine tools.

Under this IITRI project, Sculptured Surfaces Experimental (SSX) was developed as a prototype research program and successive versions were generated. And in 1972 these development efforts were assigned to the newly formed CAM-I organization, which continues to refine and improve the program.

The SSX program as it now stands cannot be implemented off-the-shelf in a production environment without considerable modification and refinement by the user. As a result, the program is still considered to be an experimental, prototype system. Yet, because of its general construction and the broad range of technology it covers, SSX is considered to be one of the most advanced sculptured surface processors available. Most other programs cannot define and reliably guide a cutter for the range of shapes of SSX. Figure 6.4 shows the undulating contours of some typical test pieces defined and machined with the SSX program.

Various versions of SSX are widely used in industry. Many companies have tailored the program to suit their individual requirements and use it as a productional tool. And some turnkey vendors have been licensed to include modified versions of SSX as part of an overall software package.

Development work on SSX has continued for more than 12 years. But experts acknowledge that an enormous amount of work remains to be done. Plans include extending the program's geometric construction and machining capabilities and providing a simpler and more user-oriented language. When the program reaches full capability, it will be integrated

Figure 6.4. Undulating contours of sculptured surface test pieces were defined and machined with the SSX processor in a CAM-I study. (Courtesy of CAM-I Inc., Arlington, Texas.)

into the totally unified CAD/CAM systems that CAM-I envisions for the future.

In sculptured surfaces technology, contours are represented as a network of patches, each of which conforms to the contour of one small surface section. Basically, the patches are like elastic unit squares that twist and distort as they stretch over the surface. Each patch can be expressed mathematically in terms of known points, vectors, and curves. Subsequently, blending and interpolation functions are used to join the patches to make a sort of "mathematical quilt" in the computer. Together, the network of patches describes nonanalytical contours that would otherwise be impossible to define mathematically.

There are many ways to insert data into the computer to construct sculptured surface patches. Basically, these methods fall into one of two broad categories: mesh-of-points and synthetic curves. The mesh-of-points definition is one of the oldest and most common methods for generating sculptured surfaces. In this approach, the user defines the surface with a mesh of coordinate points arranged on its contour.

The pattern of points in the mesh is consistently spaced so that the interconnecting splines form regularly shaped patches—typically resembling warped rectangles. If point spacing along a spline is variable, this same spacing should be followed on all splines to produce consistent patches

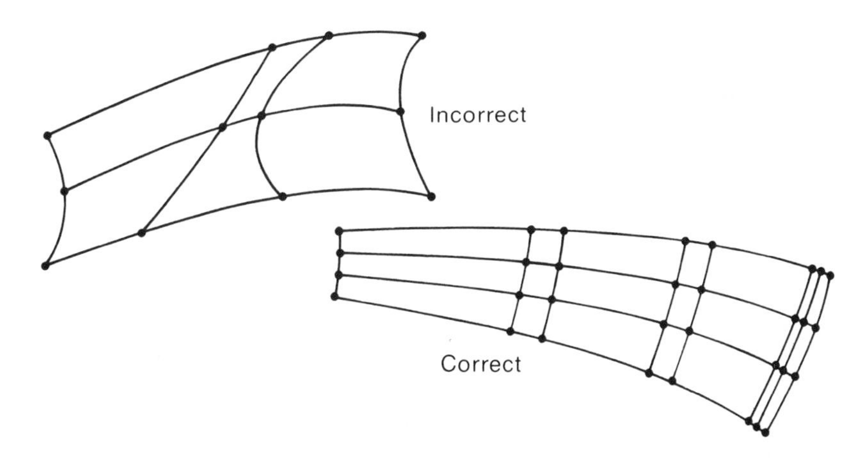

Figure 6.5. When sculptured-surface patch networks are constructed with the mesh-of-points method, input points must be properly distributed to obtain a smooth surface representation. Otherwise, the resulting odd-shaped patches may not truly represent the surface contour. (Courtesy of CAM-I Inc., Arlington, Texas.)

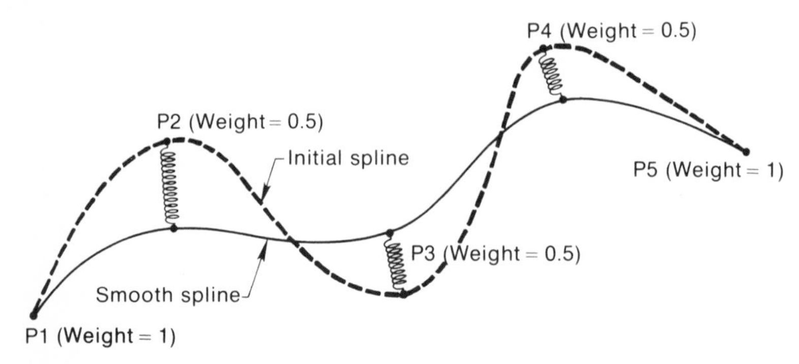

Figure 6.6. Curves for sculptured surface patches often are smoothed by a weight-fitting method in which the curve is shifted slightly from some of the input points to produce a smoother contour. The curve is visualized as an elastic beam connected by springs to fixed posts representing the input points. The beam and spring forces reach an equilibrium, removing sharp bends and smoothing the overall contour. (Courtesy of CAM-I Inc., Arlington, Texas.)

and a smooth representation. Otherwise, lack of proportionality between mating spline and cross-spline points produces odd-shaped patches, as shown in Figure 6.5, that may not truly represent the surface contour.

The computer uses a weight-fitting method, as shown in Figure 6.6, for running the splines through the mesh of points in two directions. Each point is assigned a weight indicating the value of the point in the surface definition. Also, limits are assigned to restrict the distance the patch curve can deviate from the input point. Using the weight values and limit con-

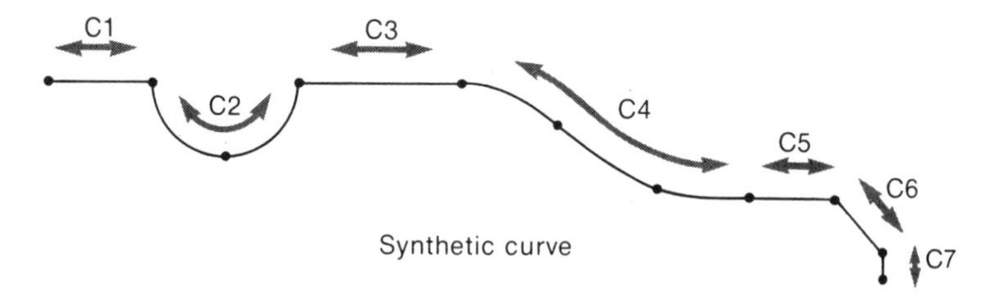

Figure 6.7. A synthetic curve is a composite line made up of individual segments. The synthetic curve shown is constructed by defining separate line segments and entering the simple command C = SCURV/COMBIN, C1, C2, C3, C4, C5, C6, C7. (Courtesy of CAM-I Inc., Arlington, Texas.)

straints, the computer generates criss-crossing spline curves that go through or near the input points. This ensures that patches have a perfect slope continuity across their shared boundaries, providing a smooth surface with no discontinuities.

The second approach to construct sculptured surface patches is with synthetic curves, which is made up of an assembly (or synthesis) of simpler curve segments such as straight lines, circles, conics, or other arcs as shown in Figure 6.7. With the synthetic curve, all the elements of a complex geometry can be combined into one entity in the computer, manipulated, and displayed as one element. As a result, highly complex shapes can be represented with comparatively small amounts of computer time and memory.

Synthetic curves may be manipulated several ways to produce a variety of surface definitions. A curve extended through space produces a tabulated cylinder. And two different curves connected with straight-line segments produce a ruled surface. These relatively simple constructions are

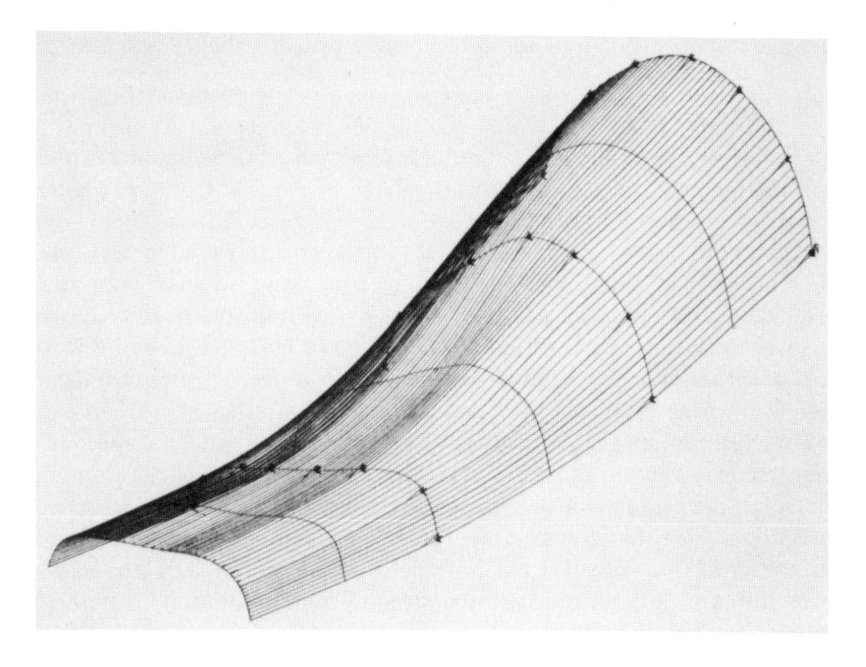

Figure 6.8. This model of an aircraft wing section was generated in a CAM-I study using the general-curves (GENCUR) sculptured-surfaces routine. (Courtesy of CAM-I Inc., Arlington, Texas.)

compatible with the traditional APT tabulated cylinder (TABCYL) and ruled surface (RLDSRF) routines. More complex constructions developed with synthetic curves are the surface of revolution and the sweep surface, which are used when the geometry contains some axial symmetry.

Completely arbitrary surfaces are represented with the general curves (or GENCUR) routine as shown in Figure 6.8. This is the most sophisticated sculptured-surface capability in which a family of splines is generated across the surface contour. Each curve differs slightly from the next and together they define every undulation in the surface.

The spline curves of the GENCUR model are produced differently from those of other sculptured surface models in which the patch network is based on user-specified corner points. With GENCUR, patches are developed with a so-called rate-of-flow concept. In this approach, cross-splines are produced by the computer to intersect the splines based on their rate of flow, or curvature. Basically, each cross-spline snakes its way across the family of splines, intersecting them at the points of similar curvature. This produces a smooth, continuous surface definition and relieves the user from coordinating points to produce properly shaped patches.

ROBOTS

A robot is a programmable manipulator arm with grippers that move and position material, parts, tools, or other objects in a CAM system. Figure 6.9 shows a typical robot performing arc-welding operations.

The simplest types are pick-and-place robots typically used in performing materials handling functions in which something is picked from one spot and placed at another. This type of robot typically moves only in two or three directions such as in and out, right and left, and up and down. And programming may be done physically with various configurations of electrical wiring or pneumatic connections.

The most common type in industry is the servo robot with servomechanisms that may alter the direction of the arm or gripper in midair. This gives the robot arm joints for greater articulated movement in five to seven directions. The robot may be connected to a programmable controller and programmed in a so-called teach mode in which an operator physically leads the arm through the required steps of an operation. This type of manual teaching is currently the most widely used robot programming technique. But it is time-consuming, and error-prone. And changes in the program usually require the entire sequence of tasks to be retaught.

The most sophisticated robots are connected to a computer from which instructions are produced and transmitted electronically. These so-called smart robots do not have to be given detailed instructions for performing tasks. Rather, the robot program automatically determines grip points and motion paths from application specifications.

Robots can manipulate objects weighing from a few ounces to over a ton and are being used extensively in industry to perform a variety of tasks. Most of these initial applications involve tasks that are too tedious or dangerous for human workers. Typical examples include loading and unloading stamping machines and handling toxic material.

As robots were refined, however, the advantage of using them to perform other tasks soon became apparent. Robots are now used in a wide range of applications including welding, parts assembly, forging, sheet-metal fabrication, and material handling. In these and other applications, the use of robots results in increased productivity, lower costs, and improved quality.

Robots typically cost $10,000 to $150,000. Yet because of tremendously

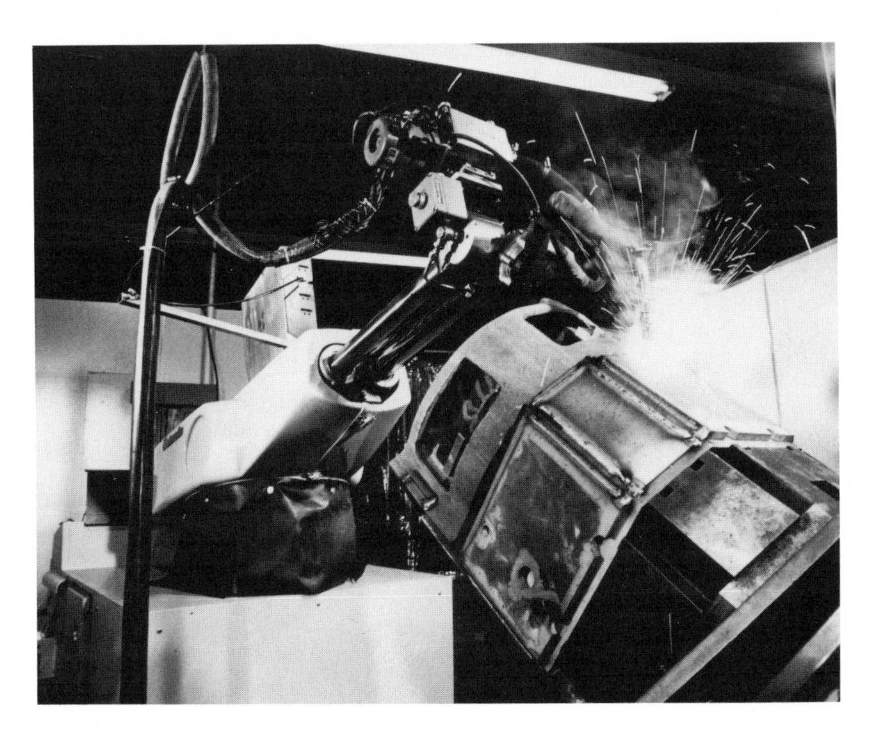

Figure 6.9. Continuous-path Unimate Robot automates welding operations for a rail vehicle. Other welding applications for robots include automobiles, heavy steel plate, tubing, and light-gage steel. (Courtesy of Unimation® Inc., Danbury, Connecticut.)

high productivity, payback periods are generally less than two years. Usually a robot can do the work of as many as six human workers. Robots cost only about $4 an hour to operate, in contrast to an average hourly labor cost of $14 for workers in the United States. As a result, tenfold reductions in costs are often possible with robots.

The automotive industry is presently one of the largest users of robots, mostly for spot welding. General Motors Corp. has 150 robot welders at work in its Lordstown facility. And Toyota and Volvo have made heavy investments in robots. One of the most advanced of these automotive welding systems is the body-framing system at Chrysler Corp. In this system, nearly 200 robots make over 98% of all spot welds on the company's K-cars.

The success of the robot in automotive work has also prompted their increasing use in general industry at companies like General Electric Co., one of the nation's largest producers of home appliances. The company has invested millions of dollars on robotics and expects to have almost 1,000 of them in operation by the end of the decade. The company is continually looking for areas to apply robots, which they find pay for themselves in cost-saving in about two years. In some typical GE applications, robots spray paint on refrigerator liners and porcelain enamel on dishwasher liners, load and unload a press that makes refrigerator liners, and spray refrigerator cabinet interiors with adhesive.

Robots also are being used increasingly in the fabrication of aircraft. One of the first applications was at General Dynamics Corp., where a robot was used as part of the ICAM program to drill mounting holes in skin panels for F-16 fighter wings. This work was started in late 1978 and productivity has reportedly doubled. Another unit brought on-line in late 1979 for drilling aluminum fuselage panels is said to have increased productivity by 400%.

So far, worker reaction to these so-called "steel-collar workers" has been favorable, even though some experts predict that as many as 20 million industrial jobs around the world could be replaced by robots in the next 20 years. For the most part, robots usually create through increased productivity more jobs than they eliminate. Even unions such as the UAW have supported robotics. So far, unions have viewed such technological progress as inevitable and essential for continued economic growth. The feeling is that the resulting economic growth more than offsets temporary job losses.

Robot capabilities continue to be refined and extended, and experts agree that robots used in combination with NC machine tools are the key elements to future automated factories. Experts envision these robots of the future evolving in two directions. In massive factories making complex products such as automobiles and aircraft, the individual robot may dissolve as a separate unit as sensors, computers, manipulators, and machine tools are combined. In smaller operations, the robot will remain a separate unit and with even larger memory and simpler programming to perform more diverse functions. For example, a single robot may feed a machine tool, move pallets and raw material, and inspect parts.

ARTIFICIAL INTELLIGENCE

Researchers are attempting to impart increasing amounts of artificial inteligence into machines with the ultimate goal being the automated factory. That is, machines are being given the capability to perform tasks that would require some level of intelligence if they were performed by humans.

Artificial intelligence does not imply any sort of consciousness or creativity, but rather a problem solving ability in which the solution is deduced from sensory input. A typical example of such decision-making capability would be the selection of parts according to shape. Future machine tools probably will have limited levels of intelligence to detect and change faulty tools, for example, or to make adaptive machining adjustment on themselves. However, most of the effort is directed toward artificial intelligence for robots.

Sensory input and the associated computer-processing capability to understand and use these inputs are essential to artificial intelligence. Some experts estimate that refined sensory robots could perform 90% of the manual tasks in industry. As with humans, one of the most versatile senses is sight. And most experts consider vision to be the key to developing the next generation of more sophisticated robots. Robots today with vision use pattern-recognition techniques in which image signals provided by TV-camera "eyes" are compared with those stored in computer memory to discriminate between various types of objects.

The robot system in Figure 6.10 can discriminate and select a single part from a number of randomly spaced parts on a pallet or conveyor belt. With pattern-recognition, distinctive parts of the object's silhouette are sensed and operated on with sophisticated software algorithms. In this manner, the data is compared with stored information to identify the part and determine its position and orientation. Other systems may use simpler

Figure 6.10. Univision Robot System is programmed with the VAL language to make responsive decisions based on visual inspection and interpretation of part disposition and orientation. (Courtesy of Unimation® Inc., Danbury, Connecticut.)

pattern-matching techniques in which the basic shape of the object is compared with that stored in computer memory.

Present vision systems are generally limited to 2D applications in which details are discriminated on the basis of silhouette. The next refinement is expected to be a more advanced 3D perception sensitive enough to guide the robot through even the most complex task. The most sophisticated systems currently under development at Stanford University and California Institute of Technology use stereo vision analogous to human sight. Another system under development at the National Bureau of Standards indentifies 3D shapes by the way a beam of light deforms as it strikes an object.

In addition to vision, the sense of touch also is essential in advanced robots. Many robots now have a rudimentary tactile sense provided by switches that detect the presence or absence of an object in the grippers. Also, grip-controlling sensors can provide force-feedback signals that enable the system to control how tightly an object is grasped. And compliant mechanisms on the grippers permit the robot to "jiggle" parts together with an action similar to that which a human would use in groping for hole locations or the proper fit of parts not precisely aligned.

Present tactile research is aimed at developing highly sensitive systems that will enable robots to handle parts more delicately and execute more complex maneuvers. One of these methods analyzes the torque applied to the robot arm joint. A more sensitive system has an array of piezoelectric material on the grippers that produces an electric current when compressed. Another system uses flexible conductive skin that produces a signal when an object forces it against an underlying array of electrodes.

Researchers in robotics are also developing voice data entry systems. These advanced systems ultimately will enable a programmer to instruct the robot verbally or an operator to enter the requirements for a specific task on the shop floor. And speech synthesizers may be incorporated to permit the robot to verbally feed data back to the programmer or operator.

The continuing goal of artificial intelligence is to integrate these sensory systems with more sophisticated computer software to enable robots to make limited decisions. Given sufficient computer power, these robots could even take appropriate actions not provided by their programmers and modify their controlling software accordingly. For example, a robot might conclude that a task could be performed more efficiently than the method given by the programmer.

Given sufficient mobility, such robots could perform the widest possible range of tasks in a CAM system. With this in mind, various methods of robot locomotion are being developed. One six-legged robot developed at Ohio State University resembles a large insect and is quite stable on rough terrain. Each leg joint is independently powered and provides position and rate feedback to the computer system which controls their movements. Other robots are being developed with two, three, and four legs. A robot being worked on by NASA has wheel-type locomotion and a sophisticated guidance system. The robot's laser-rangefinder eyes, gyrocompass, and optical-encoded odometer enable it to self-navigate in a totally alien environment.

FUTURE FACTORIES

Machine tools, robots, and the computer systems controlling them are increasing rapidly in sophistication. This direction of CAM technology will ultimately lead to what most experts are calling the automated factory, where the production process will be computer-controlled from start to finish. Theoretically, a single individual could operate such a factory. Seated at a CAD/CAM terminal of this idealized system, the operator would monitor and direct the activities of a multitude of robots and machine tools controlled by a mammoth central computer.

This idealized image of the automated factory, of course, would be too costly and complex to find widespread application in industry. More likely are future factories with various degrees of automation in each particular area. Most experts agree that human workers, robots, and machine tools will operate as teams, with each partner of the team doing what he does best.

Figure 6.11 shows this type of man-machine teamwork in a steel-forging operation for making gear teeth. Two workers monitor and direct six robots that move steel parts through the various phases of the process. Steel blanks are placed in a furnace and removed for placement in a forge. The forged part is then fed to machine tools that cut the teeth. Completing the process, a robot spray-paints the finished parts.

The potential benefits of reduced costs and increased productivity are so great that several industrial countries are pursuing vigorous research and development programs aimed at robot and NC machine-tool systems for an automated factory. In addition to the United State and Japan, West Germany, Norway, Great Britain, East Germany, The Netherlands, and Czechoslovakia are developing CAM systems that are leading to the automated factory.

Japan is in the midst of a $100 million program to have an automated factory in operation by the mid-1980s. The plant will produce metal machine components such as gears and shafts probably less than 500 kilograms in weight and one meter in outer dimensions. Every operation in the production process will be done by robots and automated machine tools, all managed by a central computer and a handful of workers. Probably 10 engineers and technicians will supervise the factory, which is expected to produce the equivalent of a conventional plant manned by 700 workers.

This ambitious project was initiated in 1977 by the Japanese government and is sponsored and coordinated by the Ministry of International

Figure 6.11. Two operators monitor and control six robots at work in a steel-forging operation making gears in a future factory. (Courtesy of Westinghouse Electric Corp., Pittsburgh, Pennsylvania.)

Trade and Industry. Work on the project is shared between several state-run research centers, universities, and private companies.

The goal of the project is an automated plant for producing low-volume parts with production runs of 300 units or less. This is one of the most technically challenging aspects of the automated factory, since low-volume production requires a flexible automation system in contrast to the rigid systems that can produce high-volume parts by the millions. Such flexible automation in which the production process is frequently changed requires the machines themselves to be flexible and the most sophisticated control systems. For example, a single complex cutting machine able to rearrange its own tools and components will be used instead of separate lathes, milling machines, and drills. Similar flexible machines will perform welding, forming, heat treating, finishing, and assembly. The key breakthrough that makes this system possible is said to be sighted robots for performing transfer and detailed assembly work.

In the United States, efforts to develop an automated factory are coordinated by the U.S. Air Force ICAM program. The technology will be developed by a joint effort between nearly 70 industrial contractors and universities with $100 million in government funding. Plans are to have an automated factory developed by the mid-1980s for making sheet-metal components for aircraft. The sheet-metal plant represents only a segment (a so-called "wedge") of the entire aerospace factory, which ultimately will include other manufacturing operations such as machining, assembly, composite material fabrication, and electronic assembly fabrication.

Sheet metal processing was selected because organizers felt technological improvements in this area would have the most immediate impact on the costs of future weapon systems. Also, much of what is learned in developing this factory will carry over into the general manufacturing done by private industry.

In the ICAM concept of the automated factory, the manufacturing process is reduced to discrete manufacturing cells made up of robots (or some other automatic material handler) and NC machine tools as shown in Figure 6.12. In a simple cell, a conveyor delivers a part, which is identified,

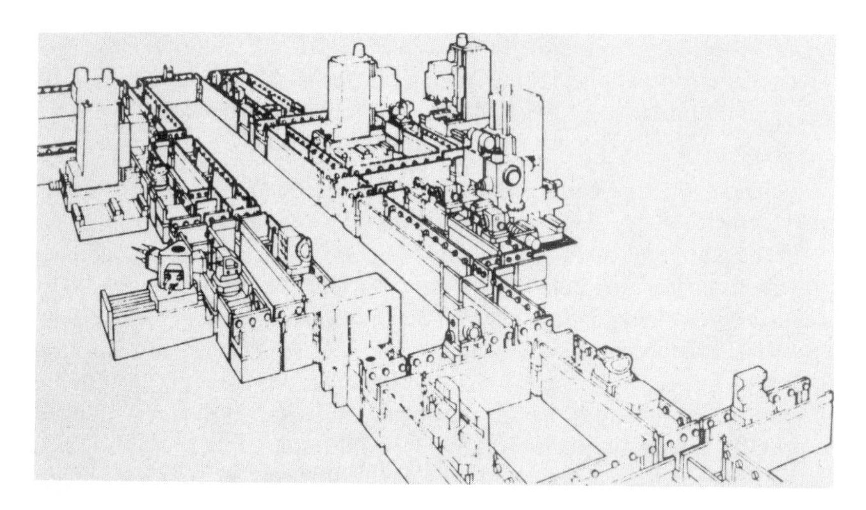

Figure 6.12. Multistation manufacturing cells such as this will evolve as the number of workstations under computer control increases. The final step in the evolution of CAM is the automated factory made up of a network of interconnected cells. (Courtesy of Westinghouse Electric Corp., Pittsburgh, Pennsylvania.)

picked up by a gripper, and presented to the machine. Or the robot may perform the required operations with specialized tools such as a welder or drill. The finished part is then passed on to another cell. Groups of cells are tied together in a so-called manufacturing center, which is in turn connected into an entire factory under the control of a master computer.

Each manufacturing cell combines two or more manufacturing processes to fabricate a family of similar parts. This is known as the group technology concept in which part families are based on the similarity of either their geometry or the manufacturing process to make them. The overall computer system controlling the factory of interconnected manufacturing (or group technology) work cells is hierarchical in nature. Built-in microcomputers control the operation of individual machines and robots, while middle-level computers coordinate the operation of one or more manufacturing cells. The overall system would be controlled by a central computer.

7
Systems at Work

Early CAD/CAM systems consisted of roomfuls of equipment costing several million dollars and required operators skilled in programming and related computer tasks. As a result, only the giant aerospace and automotive industries could afford them. And some of the most sophisticated CAD/CAM systems are still at these large companies.

But computing power has skyrocketed while size and cost have plummeted. The result has been a rapid proliferation into general industry of relatively inexpensive stand-alone equipment on which users can perform highly sophisticated design, analysis and manufacturing functions. This allows the user to reap the benefits of the computer without training in programming and related tasks. Consequently, although the most sophisticated systems remain at the large companies, many smaller companies that a few years ago could not afford CAD/CAM are now using it.

The major incentive for using CAD/CAM, of course, is increased productivity. Rising costs forced aircraft manufacturers to start implementing CAD/CAM years ago as a way to produce airplanes economically. Likewise, the automotive industry also embraced the technology as the best way to design and manufacture automobiles. CAD/CAM is now carrying over into general industry, where it is used for a wide range of products.

AEROSPACE

CAD/CAM is probably used in the aircraft industry more than in any other. And the Boeing Co. is unquestionably one of the leaders, having been heavily involved in the development and utilization of CAD/CAM since the early days of the technology. In the late 1950s, Boeing's manufacturing division used the APT language on a limited basis to describe part geometries for generating numerical control tapes. And in the early 1960s, they relied more heavily on NC in the production of machined parts for the 727 aircraft.

By the mid-1960s, Boeing was developing the 737 and had established itself as one of the largest users of NC equipment. Engineers not only used NC technology to make parts, but also to define mathematically the complex exterior surfaces of the aircraft. This led to the development of the APLFT program, which is one of the forerunners of today's sculptured surface technology.

In the early 1970s, Boeing was the first to use APT technology to generate engineering drawings. In this approach, the program was used to drive a pen plotter instead of a machine-tool cutter. One of the first major projects utilizing this innovation was the YC-14 Advanced Tactical Transport, for which a large portion of the primary wing and body structure drawings were APT-created. The experience with the YC-14 showed that APT was particularly effective in developing similar part families such as wing ribs and fuselage frames. But more important, this early work demonstrated the effectiveness of using a computer system to both design and manufacture a complex mechanical structure. These efforts laid the foundation for the development of increasingly complex systems that eventually would integrate design and manufacturing functions into unified CAD/CAM systems.

In the mid-1970s, Boeing installed interactive computer-graphics systems for production of the 747. These interactive systems enabled operators to develop part designs and the associated drawings in a real-time graphic mode. And NC was used heavily in computer-controlled manufacturing equipment such as giant riveting machines that move along the aircraft, automatically punching and reaming holes in the skin and then setting and trimming rivets.

The Boeing Co. is now extending these CAD/CAM capabilities for its newest family of 757, 767, and 777 aircraft. The 767 shown under construction in Figure 7.1 makes more extensive use of CAD/CAM than any other aircraft. About 6,000 drawings (30% of the total number for the aircraft) are expected to be produced with CAD/CAM. These drawings are for design of parts representing about 90% of the aircraft's structural weight.

Drawings for the 767 are produced using a combination of APT and interactive graphics. Basically, APT is used for surface definition of similar

Figure 7.1. Boeing makes extensive use of CAD/CAM to produce more than 6,000 drawings for 767 aircraft parts. Plans are to consolidate the sophisticated Boeing system so that engineers in different locations will have access to the same huge data base through interactive graphics terminals. (Courtesy of Control Data Corp., Minneapolis, Minnesota.)

part families while the graphics system is used for adding design details for parts such as wing panels, front and rear spars, and inspar ribs. Each member of the part family is produced by inserting key parameters into the general APT program. The resulting part geometry is then transferred to a CAD/CAM terminal and details are added. By combining all this information into a common data base, the entire aircraft may be assembled in the computer to check for clearance and fit. For example, landing gear action may be studied, as well as wire and pipe routing through the aircraft sections.

Plans are to consolidate the Boeing computer services so that all engineers will have access to the same huge data base through interactive graphics terminals. This will allow engineers at various location in the United States and overseas to communicate design and analysis data instantaneously through graphics terminals instead of with drawings. Consequently, time will be drastically reduced for solving problems, in-

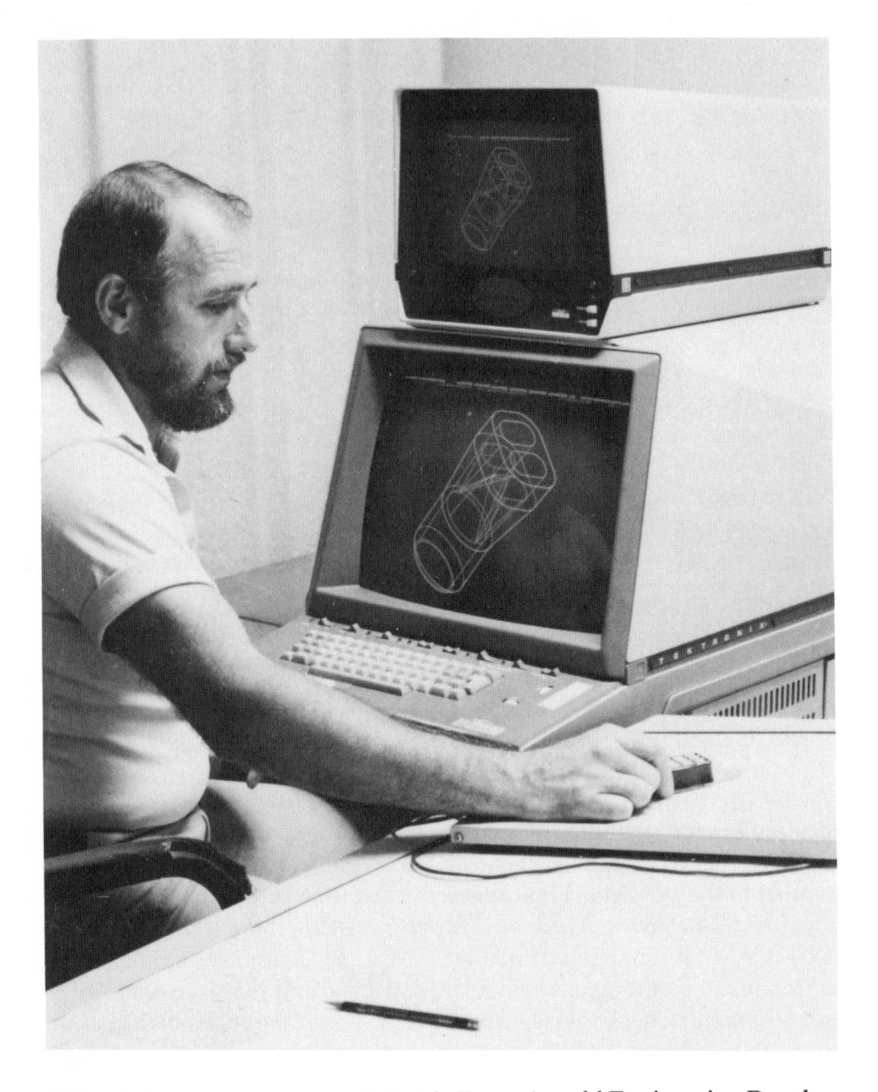

Figure 7.2. Designer at the U.S. Air Force Arnold Engineering Development Center uses an electronic cursor to create a design of a prototype mechanical component. Using the CAD/CAM terminal, the designer can generate three times as many mechanical drawings of better quality than he could produce manually. (Courtesy of Arnold Engineering Development Center, U.S. Air Force, Arnold AFS, Tennessee.)

tegrating and completing design and analysis, and communicating project progress.

All the major aircraft producers as well as the U.S. Air Force presently are expanding their use of CAD/CAM. The most concentrated Air Force effort is, of course, the ICAM program with the ultimate goal of the automated factory for the production of aircraft. However, the use of CAD/CAM also is permeating other Air Force applications. At Arnold Engineering Development Center in Tennessee, CAD/CAM systems such as the one in Figure 7.2 are used to produce drawings for special project applications, including the design of wind-tunnel hardware and aerodynamic models.

Hundreds of engineering drawings, including complete floor plans for most of the engineering center buildings, are stored in the Arnold system for future use. The present system is limited to 2D geometry, but plans are to expand to a full 3D system. When fully implemented, this more advanced system will be used to manufacture complex aerospace components on DNC systems and to support prototype Air Force test projects.

Manufacturers of light aircraft also are using CAD/CAM increasingly. The Piper Co. most recently used the MCAUTO CADD system for the creation of engineering and tooling drawings for the Cheyenne III in Figure 7.3. Design of aircraft parts, as shown in Figure 7.4, provides considerable savings in design time and reduction of errors. Piper also has discovered that CAD/CAM improves communications between its separate operat-

Figure 7.3. Piper's 11-seat turboprop Cheyenne III aircraft was designed with the MCAUTO CADD system. (Courtesy of McDonnell Douglas Automation Co., St. Louis, Missouri.)

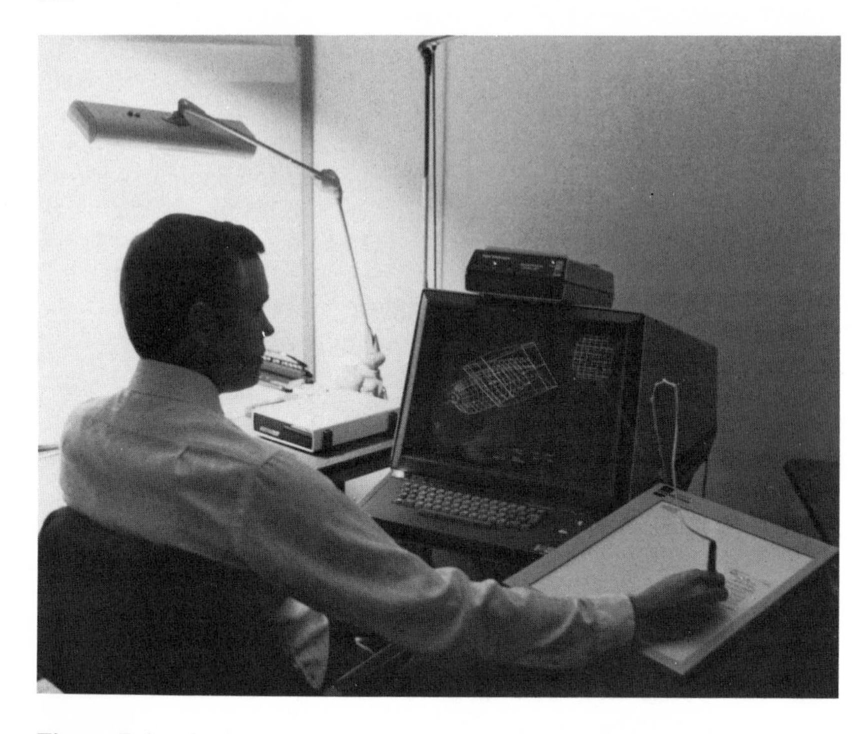

Figure 7.4. Senior engineer at Piper's Advanced Engineering Department designs a wing section and nacelle on the CADD system for the Cheyenne III aircraft. (Courtesy of McDonnell Douglas Automation Co., St. Louis, Missouri.)

ing divisions in Florida, Pennsylvania, and California. With design data transmitted to graphics terminals over telephone lines, engineers at any of these locations can simultaneously review any given design.

Piper is also using FASTCUT (a subsystem of CADD) to produce NC instructions for punch presses and multi-axis milling and drilling machines. FASTCUT interactively creates a tape from the CADD geometry to control scribing, drilling, reaming, milling, and punching operations. As a result, Piper can design an aircraft part with the CADD system and create the NC instructions to make it with the FASTCUT subsystem.

AUTOMOTIVE

The automotive industry has been particularly aggressive in applying CAD/CAM. Computer technology allows Detroit engineers to develop new designs for increasingly stringent safety requirements and to reduce weight for fuel-economy standards. In addition, product development time and costs are reduced for automobiles designed with CAD/CAM.

Basically, computer technology enables engineers to more effectively manage the myriad of interrelated variables inherent to automobile

Figure 7.5. Computer model of an Aspen/Volare is displayed on an Imlac terminal. This sort of concept surfacing is used to visualize the shape of the overall vehicle. Detailed component design comes later. (Courtesy of Imlac Corp., Needham, Massachusetts.)

design. Industry observers have said that the safety, environmental, and fuel-economy requirements mandated by the federal government are so stringent and have been imposed so rapidly that it would be virtually impossible for humans alone to make the engineering changes fast enough. They say that computers are a necessity for the automotive designs to remain in compliance with federal standards.

In the early stages of design, the interactive graphics capability of CAD/CAM can be used to visualize the overall shape of the automobile. In this so-called concept surfacing, the contour of the entire vehicle is developed and evaluated as shown in Figure 7.5. Interactive graphics also

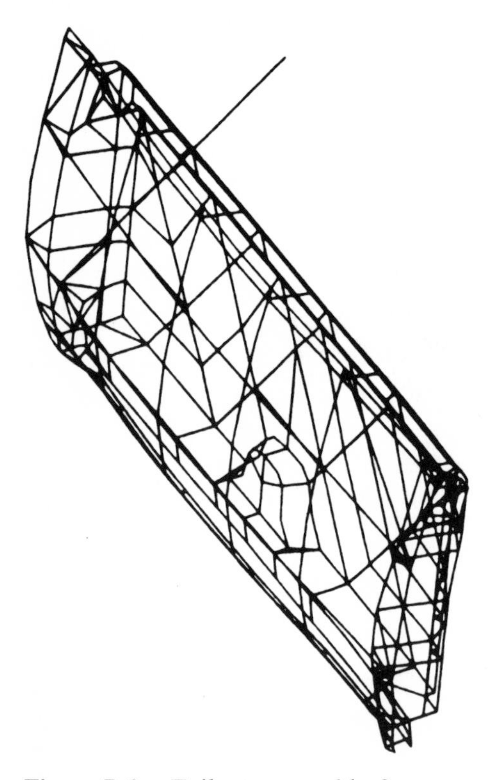

Figure 7.6. Tailgate assembly from a station wagon was modeled with the SUPERTAB software package. Analysis of the geometry enabled designers to use less material in the part without introducing high stress concentrations. (Courtesy of Structural Dynamics Research Corp., Milford, Ohio.)

Figure 7.7. Designer creates a model of an automobile transaxle assembly with keyboard and electronic tablet. The resulting data base then may be accessed by other engineering areas to perform structural analysis, generate drawings, or produce NC tapes. (Courtesy of Computervision Corp., Bedford, Massachusetts.)

is used in the detailed design and analysis of components such as the tailgate or the transaxle assembly in Figures 7.6 and 7.7. And in the case of moving components, kinematic capabilities may be employed to animate the mechanism on the screen as in Figure 7.8. The resulting data base that is compiled for the design of all these individual components than may be stored in the computer and accessed by other engineering areas to perform structural analysis, generate drawings, or produce NC tapes.

But mechanical components are not the only elements modeled with the computer. One innovative use of CAD/CAM technology at Chrysler Corp. is a manikin called "Cyberman," a montage of human measure-

Figure 7.8. Operator studies the movement of an automobile piston assembly on a Calma DDM system. (Courtesy of Calma Co., Santa Clara, California.)

ments illustrated in Figure 7.9. This computer model is used to evaluate the packaging of interior features such as seats and armrests. The computer-based manikin enables engineers to predict the positions of passengers while the vehicle is in the early stages of design, saving the expense of creating physical mockups. The model can be programmed into a number of positions and displayed on a graphics terminal as a stick figure or a complete wire-frame outline.

System simulation is an important aspect of CAD/CAM applied extensively in the automotive industry. In this approach, an analytical model of the vehicle responds to loads and driving conditions in much the same manner as the real car. The system model in the computer represents structural characteristics of components from tires and shock absorbers to the frame and sheet metal. Additional data is entered with the system model representing external loads such as tire impact with a curb the vehicle may encounter during operation.

The computer manipulates these sets of data and provides a prediction of total vehicle response in the form of an animated mode shape showing structural deformation. Actual displacements are relatively small and rapid; so these animated modes shapes such as in Figure 7.10 show distortion in slow motion with amplitudes exaggerated for clarity.

Based on these animated mode shapes, the automobile design may be modified and resimulated until it performs satisfactorily. In this way, the design is developed and modified in the computer rather than with hardware prototypes. Consequently, CAD/CAM designs generally are closer to an optimum because numerous alternatives can be simulated and the best one refined. In contrast, conventional build-and-test procedures rely heavily on physical tests of the final vehicle configuration. These manual designs may be less than optimal because quick-fix changes often are made late in product development to correct structural deficiencies uncovered only after the prototype fails. The utility of CAD/CAM technology in automotive design was demonstrated when General Motors downsized their full-sized B cars using computer simulation. The technique enabled

Figure 7.9. Cyberman is a computer model of the human body used by automotive engineers to position interior components according to human measurements. (Courtesy of Control Data Corp., Minneapolis, Minnesota.)

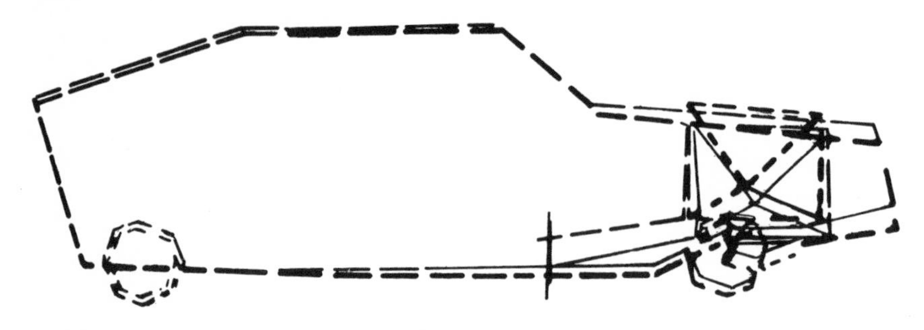

Figure 7.10. Animated mode shape of automobile pinpoints the source of excessive vibration—the engine bouncing out-of-phase with the front end. Ride quality was smoothed by repositioning the engine mounts and reducing the stiffness. (Courtesy of Structural Dynamics Research Corp., Milford, Ohio.)

engineers to reduce vehicle weight by about 500 pounds while at the same time improving handling and ride quality. Another benefit of computer simulation in automobile work is reduced product development time and cost. Design of the Cadillac Seville, for example, relied heavily on computer simulation and consequently a refined prototype of the car was developed six months sooner than was customary.

Finite element analysis is another CAD/CAM technique used extensively in the automobile industry to indicate areas of high stress where parts such as fenders are most likely to fail. Using this method, analysts can quickly evaluate the effect of modifying the contour or changing materials in a part. Chrysler Corp., for example, plans to reduce the average weight of their entire fleet by 1,300 pounds by using the finite element method to evaluate the use of lightweight materials such as plastics, graphite, and high-strength steel. Likewise, Ford Motor Co. is using extensively, the finite-element method and related technology, and the Escort/Lynx models are the first lines introduced by the company in which CAD was used in all phases of development.

Finite-element analysis also is used to evaluate the crashworthiness of automobiles. In the early 1960s, barrier-impact testing was the only method for such an evaluation. Later in the decade, computer simulation based on test data was used to study collision dynamics of the vehicle and occupants, and crash simulation has become a reliable automotive design tool. But the method still requires experimental inputs to represent the large plastic deformations in the structure.

Analysts are now attempting to develop refined finite-element techniques as a more convenient way to study vehicle crashworthiness. The method is particularly suitable to the problem because of the irregular geometry of the automobile and the huge amount of data to be correlated. The ultimate goal is to determine the dynamic forces and displacements for an entire vehicle with a finite-element model.

As yet, developers have not modeled an entire automobile accurately enough to predict reliably these forces and displacements. This is because of the complexity of the automobile structure, which consists of shells, beams, corrugated panels, irregular bars and thin-walled beams and columns all subjected to large plastic deformations. However, efforts are continuing to develop accurate models for the frame, sheet-metal body, bumpers, and other components. And developers hope to blend these individual areas into a composite modeling technique for depicting the entire vehicle.

GENERAL INDUSTRY

Although the aerospace and automotive giants are still the largest users of CAD/CAM, its use is rapidly spreading into many diverse areas of general industry. Many of the same techniques used in designing and manufacturing aircraft and automobiles transfer directly to other types of mobile equipment such as construction equipment and agricultural machinery. As a result, companies such as International Harvester Co. and Caterpiller Tractor Co. are committing large amounts of capital to CAD/CAM.

In addition, the military is also investing heavily in CAD/CAM. The U.S. Navy is spending over $10 million on a sophisticated interactive graphics system for designing ships. Models of various ship types will be stored in the computer system. From this data base, detailed structural designs can be produced and modified. Designers may also enter data on wind, wave motion, and other environmental conditions to test the seaworthiness of the ship,

CAD/CAM is also being applied to other complex mechanical systems such as machine tools. Bendix Machine Tool Corp. has a system for designing gear trains that connect motor drives to multiple spindle heads. The CAD/CAM system aids the designer in determining the proper gear sizes and gear combinations to provide the required spindle output speed. The system also is used to determine whether the gears will fit into the gearbox without interference. In addition, the CAD/CAM system has the capability to produce NC tapes for machining the gearbox.

Bendix also has a program that allows comparison of cost and efficiency for various transfer line designs. The software is similar to that used at Ford and GM to determine the efficiency of automobile assembly systems. With this system, various alternatives for the transfer line are entered, and the system displays the corresponding efficiency of operation. Factors considered by the computer include machine type, pallets per machine, parts per pallet, pallet size, workstation base length, cycle time, tool change frequency, and transfer distance and transfer time.

CAD/CAM also is used to simulate the operation of machine tools. This capability is useful in determining tool paths and other variables. And modal analysis techniques such as in Figure 7.11 can be used to determine how the machine will vibrate during operation. The design can then be modified and resimulated until satisfactory operation is achieved. These capabilities permit an optimal design to be reached quickly, thus reducing product lead-time dramatically. This advantage of CAD/CAM is often one of the most important considerations to a company using CAD/CAM.

At Hughes Tool Co., the time it takes to design oil-well rock-drilling bits is being cut in half with an Applicon CAD system shown in Figure 7.12. This sort of lead-time reduction is important in oilfield exploration, where

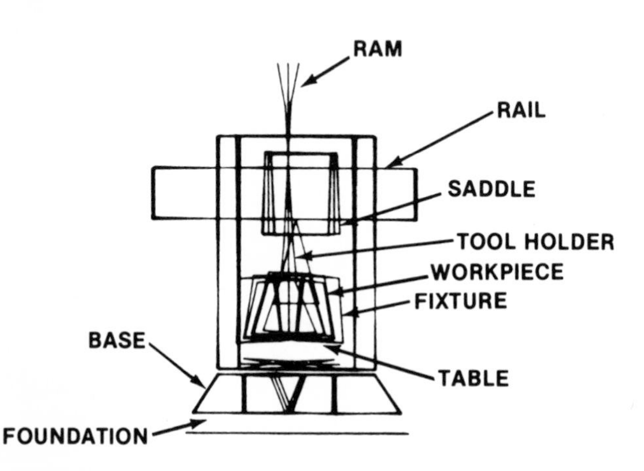

Figure 7.11. Animated mode shape of machine tool shows the lateral movement of the workpiece and fixture at 52 Hz. Such mode shapes enable designers to locate excessive vibrations before the equipment is built, and modifications can be made in early design. (Courtesy of Structural Dynamics Research Corp., Milford, Ohio.)

Figure 7.12. Designer at Hughes Tool Co. constructs model of rock bit at an Applicon graphics terminal. The resulting data base is used to produce drawings on a Calcomp plotter. (Courtesy of Applicon Inc., Burlington, Massachusetts.)

bits often are custom-designed on short notice to handle unusual drilling problems encountered in the field.

To attain this increase in productivity, the system is used with the "family of parts" design approach. A variety of standard bit designs are stored in the system and classified according to the size and type of rock they can cut. To develop a customized bit, the user selects one of these base designs from computer memory, views it on the terminal screen, and modifies it according to the particular application. Consequently, the time-consuming task of manually redrawing each new bit design from scratch is eliminated.

To increase their productivity further, Hughes customized the standard Applicon function menu for rock-bit design. For example, geometries of parts, such as bearings and cutters frequently used in rock bits, were added to the standard drawing-command menu. With simple keyboard commands, the user scales and positions these elements on the design instead of producing the configuration from scratch time after time. Characteristic data on bodies, teeth, materials, and bearings used in bits is also stored in the system. This enables engineers to select components at the keyboard instead of with stacks of design manuals. After the design is complete, the resulting data base is used to produce engineering drawings on a plotter. In the future, an expanded system is expected to use the data base for creating finite-element models and NC instructions.

The analytical capabilities of CAD/CAM are applied to structures that formerly were not considered suitable for rigorous analysis. Dust collectors, for example, are now being analyzed at Torit Div. of Donaldson Co. with the finite-element method as shown in Figure 7.13. Finite element analysis determines if a newly designed dust collector will meet standards before it is built, eliminating the time and cost of building prototypes.

The first product Torit analyzed by computer was a 17-foot-high filter cartridge collector with a 3-foot module cantilevered over the base structure. The UNISTUC program was used to construct the finite element model of the structure. The model was then analyzed with the ANSYS program, which is used frequently to verify the design of buildings and other large structures for certification. Loads were applied to the collector model to simulate wind, seismic vibrations, and other forces. The integrity of the structure was then verified by analyzing the stress, deflection, and reaction-force distribution on the collector's channel frame that ties the upper cabinet to the hopper.

One of the largest companies in general industry using CAD/CAM extensively is General Electric Co. One of every ten GE engineers now has ac-

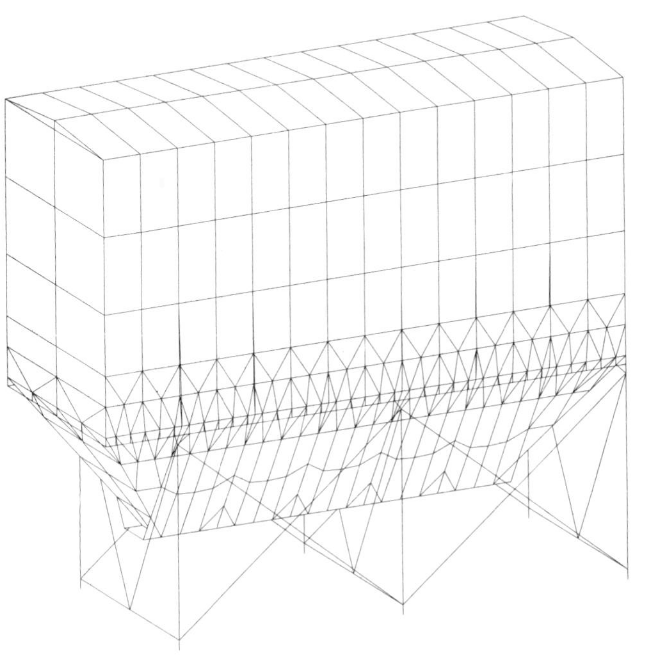

Figure 7.13. A dust collector is analyzed with a finite-element model. Loads applied to the model simulate wind, seismic, and other forces to evaluate the structure before it is built. (Courtesy of Torit Division, Donaldson Co., Minneapolis, Minnesota.)

cess to an interactive graphics system, and the company expects to increase the number of terminals tenfold over the next ten years. Presently there are about 800 interactive graphics applications at GE, most of which are for designing appliances and related products.

GE's investment in CAD/CAM is high. Out of a total GE capital investment of $1.4 billion in 1980, one-fourth to one-third will go toward systems and equipment with a significant content of CAD/CAM capability. GE is currently the biggest single corporate user of interactive graphics equipment, and the largest customer of turnkey vendors like Computervision and Applicon.

The company uses CAD/CAM for a variety of functions in its products. With finite element analysis, for example, cooling of molded plastic parts is simulated to determine if cooling is uniform. The mold design can then be changed and resimulated until a satisfactory distribution of heat is reached. This is an improvement over the traditional method of building a mold and modifying it in a trial-and-error process. And because GE makes more than 130 million pounds of plastic parts a year, this capability is expected to save over $100 million annually. Furthermore, they expect the same finite element technique to be used for castings, forgings, and extrusion.

8

Cooperative Efforts

Advancing and integrating the many diverse aspects of CAD/CAM technology is too complex a task to be attempted solely by one company, or even an entire industry. However, the potential benefits of CAD/CAM are too great to afford waiting for full integration to happen by itself. As a result, private industry, governments, and universities around the world are cooperating in a joint effort to advance individual aspects of CAD/CAM and fuse them into a unified technology.

Many of these individual segments of CAD/CAM are new, emerging only within the last few years. Others are relatively mature technologies that have been only recently refined and enhanced. The greatest challenge facing CAD/CAM today is developing integrated systems in which information can flow freely between all these diverse segments.

CAM-I

Computer-Aided Manufacturing International, Inc. (CAM-I) is a non-profit organization created in 1972 by a group of industrial companies, educational institutions, and government agencies that pooled their resources to solve common generic CAM problems. CAM-I currently has more than 500 technical representatives from 130 organizations throughout the world. Acting as a unified group, the member organizations develop and advance computer technology in design, analysis, and manufacturing more effectively than if they acted separately. By spreading out the cost of research and development and sharing in the resulting technology, CAM-I members have access to advanced CAM technology that would otherwise be prohibitively expensive.

Essentially, CAM-I provides a conduit for transferring technical information between the members. And the organization acts as a forum for discussing and implementing an overall direction for CAD/CAM development. CAM-I members develop and execute long-range technical planning for the application of computers. Moreover, the organization develops concepts and reduces them to specifications and prototype software to demonstrate and justify industrial application.

The overall activities of the organization are guided by technical committees to which member organizations appoint representatives. These are the Advanced Technical Planning, Standards, and Education-Industry committees.

The Advanced Technical Planning Committee basically defines the technical directions open to CAM-I by evaluating the impact of computers and manufacturing and identifying technical areas that require further development. The Standards Committee develops standards for computer languages, data structures, and other computer techniques. And the Education-Industry Committee encourages cooperation and interaction between the educational community, industry, and government CAM teaching and research.

These committees guide CAM-I activities in specific areas of CAD/CAM technology through projects manned by representatives from each of the sponsoring member organizations. The project teams hold periodic meetings, determine system specifications, and develop prototype software in each of their respective areas. CAM-I presently has six active projects: Framework, Geometric Modeling, Process Planning, Advanced Numerical Control, Sculptured Surfaces, and Factory Management.

The Framework Project basically ties together the development work going on in all the diverse areas of CAM, providing an overall strategy for such technical development that is essential for a concerted effort. To coordinate activity between the various CAM functions, the project is developing a system framework which will serve as a vehicle for cooperation between the many diverse international groups working in various CAM areas. The framework will also provide a common basis for communicating the resulting technical developments to general industry.

The Geometric Modeling Project is developing a generalized geometric modeling system to adequately describe arbitrary 3D shapes for design and manufacturing. To develop a prototype system with this capability, the project is studying and evaluating existing geometric modeling techniques. Also, a consistent set of terminology is being developed for geometric modeling. And work is underway to determine how to best integrate such a generalized geometric modeling system with other CAM functions.

The Process Planning Project is developing an advanced system capable of automatically generating manufacturing process plans and work instructions from product description data. The Experimental Planning System (XPS) is expected to ultimately operate in both the retrieval and generative modes. The retrieval mode uses group technology to modify plans for similar parts, and the generative mode uses planning logic in the system to automatically generate process plans from geometric and other part description data.

The Advanced Numerical Control Project is developing a fully computerized NC processor capable of generating optimum tool paths and automating machine technology decisions. Ultimately, the processor will use generative programming techniques to recognize the solid part model, identify the material to be removed, automatically select the appropriate tools and machining operations, and determine optimum tool motion and speed. As a first step in developing a prototype processor, the project has compiled a set of system specifications and conceptual design of the processor is complete.

The Sculptured Surfaces Project is providing research and development of computer-aided methods to define and manufacture complex surfaces. This will ultimately enable the mathematical representation and NC machining of difficult, contoured geometries. The project has made advancements in developing a prototype Sculptured Surfaces Processor (SSX) with these capabilities. The project is also working toward extending the sculptured surfaces geometry with the solid 3D modeling capabilities of the Geometric Modeling Project.

The Factory Management Project is developing concepts for a system to control shop-floor operations and provide for effective communication between the factory planning and operating functions. This system ultimately will be a hierarchically distributed, real-time information system. For developing this system, the project is presently working on defining the levels of factory control and reporting and determining the control and decision-aiding mechanisms such as predictions of the effects of proposed actions for each of these levels.

The relationship between the CAM-I projects is shown in Figure 8.1, which is a diagram from the long-range plan of the Advanced Technical Planning Committee depicting how information ultimately will flow in a totally integrated CAD/CAM system. In this diagram, the CAM-I projects are denoted by boxes with their initials. Cylinders indicate information services which store, retrieve, transfer, analyze, and process information in the system. The four information services are: Product, Process, Processor, and Production.

The Product Information Service is concerned with the Geometric Modeling Project and deals with the physical products of manufacturing, their constituent parts, and their relation to one another. The Processor Information Service is concerned with the plant, equipment, and tools to make the products and is a part of the Process Planning, Advanced NC, and Sculptured Surfaces projects. Also a part of these projects is the Process Information Service, which links the Product and Production Information services and serves to translate the physical form of the product as a series of actions by men and machines. The Production Information Service is the major target of the Factory Management Project and deals with production plans, schedules, and shop-floor events.

This type of organization and long-range planning is essential to coor-

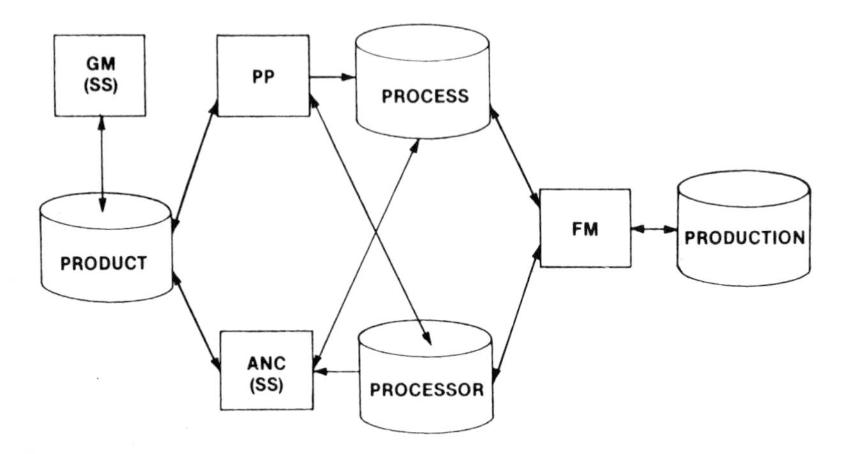

Figure 8.1. CAM-I's long-range plan calls for an ultimate CAD/CAM system consisting of four information services: Product, Process, Processor, and Production. These services support manufacturing activities through information storage, retrieval, communications, analysis, and processing. Associated with each service are specific CAM-I projects: Geometric Modeling (GM), Sculptured Surfaces (SS), Process Planning (PP), Advanced Numerical Control (ANC), and Factory Management (FM). (Courtesy of CAM-I Inc., Arlington, Texas.)

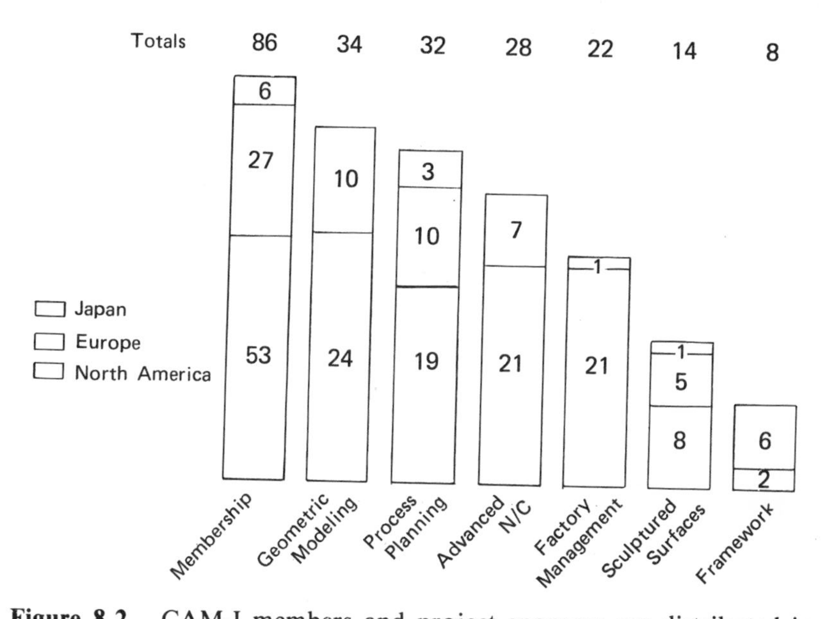

Figure 8.2. CAM-I members and project sponsors are distributed in North America, Europe, and Japan. (Courtesy of CAM-I Inc., Arlington, Texas.)

dinated efforts in CAD/CAM and is considered to be one of the most important benefits of CAM-I involvement. Consequently, international membership and project sponsorship continue to expand as shown in Figure 8.2. CAM-I members are currently distributed in Western Europe, North America, Japan, and Australia. About 70% of CAM-I's project sponsorship funds came from the United States, 27% from Europe, and 4% from Japan. The United States has been the dominate region since the formation of CAM-I, but European involvement in CAM-I has been growing rapidly. Consequently, CAM-I is currently adapting many of its operating procedures to promote this international cooperation.

IPAD

The Integrated Program for Aerospace-Vehicle Development (IPAD) is a NASA-sponsored project underway since 1976. The project has grown since its beginning into a $15-million effort presently funded at $4-million a year. Basically, the project is aimed at developing a software program for integrating CAD functions and for developing efficient ways to handle the huge quantities of data in such systems.

One of the major concepts is to store and transmit technical data via the CAD/CAM system rather than with engineering drawings, which will be used primarily to display results. The focus involves handling data in four major areas: drafting, material properties, geometric characteristics, and analysis. The ultimate goal of IPAD is to increase productivity in the aerospace industry. But the resulting technology also is expected to benefit other areas such as civil engineering and automotive design.

The IPAD project will basically provide a software framework within which all the separate functions of CAD will be able to operate in an integrated manner. Functions such as finite-element analysis and geometric modeling already are interfaced in many systems. But one IPAD spokesman characterizes them as "quick and dirty fixes rather than truly integrated systems."

The IPAD program is being developed by the Boeing Co. under contract to NASA. And development of the program is being coordinated with the U.S. Air Force Integrated Computer-Aided Manufacturing (ICAM) project. As a result, IPAD provides for interfacing design and manufacturing data within a company as well as with other companies. Developers are not trying to create production-level software for use off-the-shelf in a plant. Rather, they are attempting to develop a prototype system to demonstrate the technology on which they hope others will build.

Aerospace companies will be given the IPAD system in increments for use on their own computer systems. As a result, IPAD will augment rather than replace the existing operating systems. And the gradual transition from current computing techniques to IPAD integrated techniques will be at a pace selected by each company for its own implementation.

The capacity of the IPAD system is expected to be limited only by the user's hardware configuration. The system is intended for use in a distributed computing network with one or more central host computers and many remote computing systems. Essentially, IPAD software will function as a third-generation computer system in use today by large aerospace companies. The program is envisioned as having four major software elements as shown in Figure 8.3. These are the executive, utility, data management, and interface software.

Essentially, the executive and interface software serves to coordinate IPAD communications. Executive software acts as an interface connect-

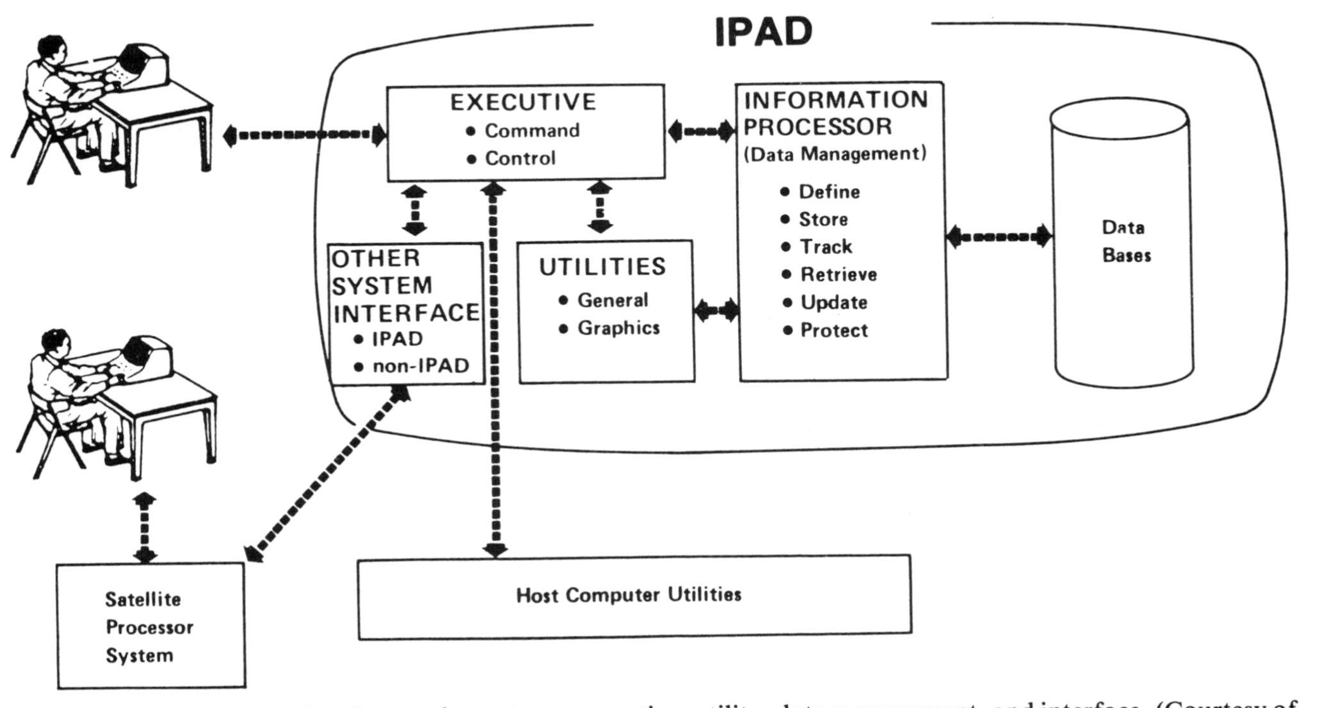

Figure 8.3. Major IPAD software elements are executive, utility, data management, and interface. (Courtesy of IPAD, National Aeronautics and Space Administration, Hampton, Virginia.)

ing the many interactive graphics terminals and other software packages with the host computer. And interface software provides for communications to computer systems outside of IPAD.

Utility software packages perform routine display and data-manipulation functions. The utility programs provide state-of-the-art capabilities in areas such as graphics, drafting, and finite-element modeling. These programs are supported by IPAD so that in a unified CAD/CAM system the design may be created, analyzed, and released to manufacturing in an integrated design environment. The geometry is stored in an IPAD standard geometry format and easily communicated to various areas of CAD/CAM. In addition, IPAD also provides access to an application program library of user-produced programs. These are programs developed by the company to apply CAD/CAM technology to its specific product line.

Data management software (also called the information processor) provides for storage, tracking, and retrieving large quantities of data maintained on multiple storage devices. This sort of data management is considered the major function of IPAD. The amounts of data to be handled in an integrated computer-aided engineering system with up to 1,000 terminals in simultaneous use is expected to be from 100 billion to a trillion bits.

The IPAD system stores and communicates many types of data. This includes business data handled in a manner similar to many commercial data base management systems. In addition to this, however, the IPAD system uses an enhanced version of present commercial data-base technology to handle engineering data, which has more complex relationships than business data and includes elements such as coefficients and polynomial expressions.

Some of the major data types stored in IPAD include: design/analysis, numerical definition, part control, and support. Design/analysis data includes dimensional units, physical constants, and weights and loads. Numerical definition is surface geometry, detailed part geometry, and mathematical modeling. Part control includes part numbers, classification codes, and heat treatment. Support includes spare part data, reliability, and operational and maintenance manuals.

One important aspect of IPAD is that its work is strongly connected with an Industrial Technical Advisory Board (ITAB) formed soon after IPAD was first started. ITAB provides industry an opportunity to influence the course of the program development. The board reviews planning and technical documents, critiques key development decisions, ranks

IPAD requirements, identifies demonstration programs, and considers the formation of an IPAD user group.

The advisory board is made up of voting members and observers which meet periodically. The voting members which control IPAD's direction include Boeing and ten other major airframe builders, engine manufacturers General Electric Co. and Pratt & Whitney, and four computer vendors: IBM Corp., Control Data Corp., Digital Equipment Corp., and Sperry Univac. Nonvoting observers include six more aerospace firms, equipment manufacturers such as TRW Inc. and Westinghouse, automobile manufacturers Ford Motor Co. and GM Corp., ten software and interactive graphics firms, ICAM, six universities and research institutes, the national Bureau of Standards, and the Naval Air Systems Command.

ICAM

Integrated Computer-Aided Manufacturing (ICAM) is a U.S. Air Force project aimed at developing a master program to coordinate all the sophisticated computer-aided design and manufacturing functions now used piecemeal by industry. ICAM is attempting to ultimately organize every step of the manufacturing process around computer automation. The program is expected to integrate such diverse engineering functions as design, analysis, fabrication, materials handling, and inspection. Ultimately, ICAM users will be able to design a part and, at the same time, evaluate its performance, plan it fabrication, and determine its cost.

The ICAM program is funded by the U.S. Department of Defense and managed by the U.S. Air Force Materials Laboratory, which pioneered the development of NC machine tools and the APT programming language in the 1950s. The $100 million program was started in 1978 primarily for increasing productivity in the aerospace industry, but the technology is also applicable to other manufacturing industries.

Basically, ICAM provides "seed money" to develop the technology. The U.S. Air Force is supplying this risk capital as a large CAD/CAM user with great potential to gain from these developments. With government funding, work is being undertaken in industry and universities, and the resulting technical developments are made available for industry to apply freely.

A major objective of ICAM is to remove the slash from CAD/CAM, thus eliminating the boundary and creating a unified technology. To ensure this happens, ICAM is working closely with the IPAD program. Both ICAM and IPAD programs are planned to be made available to design and manufacturing and to have a common language for geometric descriptions.

ICAM program developers envision the total manufacturing process as being comprised of distinct areas called wedges. Each wedge starts with a specific shop-floor process and widens to include more general functions. Hardware and software for each wedge will be developed and integrated with other wedges until the master ICAM system is completed sometime in the 1990s. The first shop-floor process selected for the ICAM development is sheet-metal processing. Planners feel that improving operations in this area will significantly reduce manufacturing costs. The next wedges to be tackled will be sheet-metal assembly and electronic device manufacturing, which will be under Army sponsorship. The next probable wedge after that will be composite materials fabrication.

The ICAM logo shown in Figure 8.4 represents the major areas ICAM covers. In the center is the architecture (or model) of manufacturing, around which the entire program is structured. The ring around the center symbolizes the computer system which handles the integrated ICAM data base. The next outward ring represents the systems required for sound manufacturing decision making: group technology, simulation, mathematics, operations research, and advanced planning methods. The outer ring represents the shop-floor containing both systems (fabrication assembly, test, and materials handling) and support activities (control, external functions, and design).

A critical concept in the development of the ICAM program is the reduction of manufacturing processes into discrete hierarchies. These are categorized as process, station, cell, center, and factory stages. Each of these stages has its own software and hardware as shown in Figure 8.5. Orderly development of ICAM requires each stage to be properly sequenced and coordinated in relation to the manufacturing process. ICAM is expected to be integrated into the lower stages of process and station which depend heavily on hardware developments. As efforts in these lower hierarchies are demonstrated, higher stages depending more on software developments will be gradually integrated into the program.

To establish the hardware and software for the sheet-metal center, ICAM initiated a robotics project for drilling and routing aircraft sheet-metal panels. In this project, a Cincinnati Milacron T^3 Industrial Arm was upgraded from a simple work-station to what ICAM defines as a work cell. A work cell combines two or more manufacturig processes to create a family of similar parts. Materials handling systems connect these cells to create a so-called workcenter.

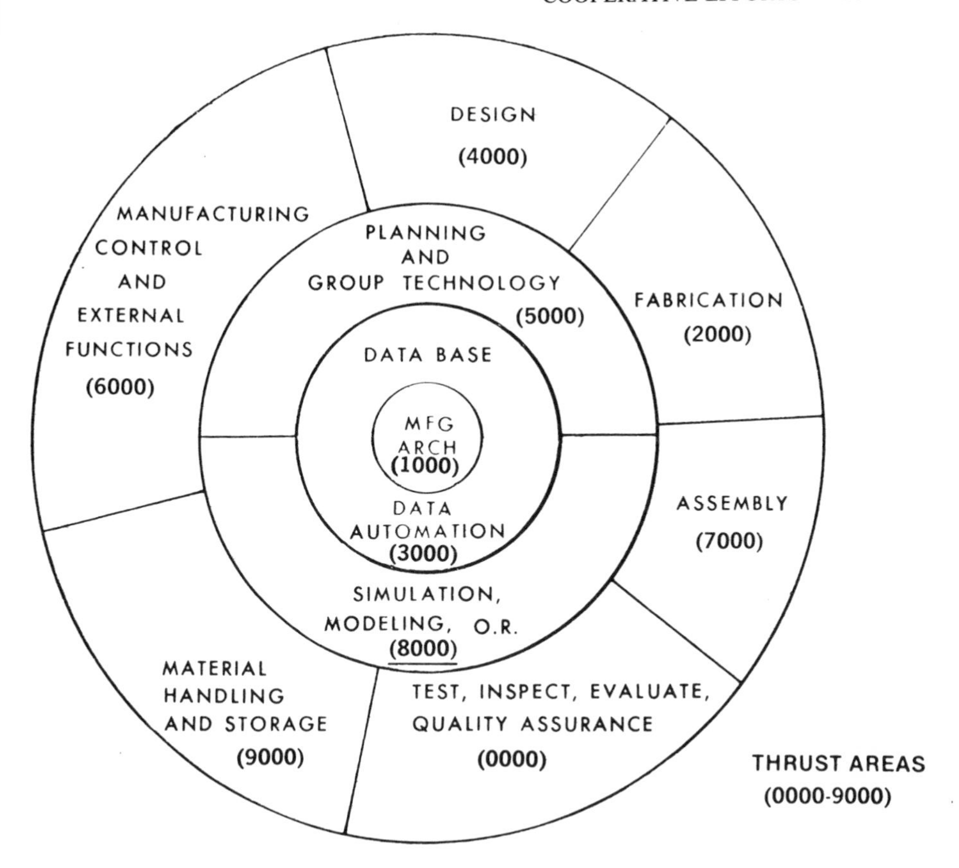

Figure 8.4. The ICAM logo is a graphic representation of its major thrust areas, shown here identified with four-digit numbers. In this concentric pattern, there is a gradual transition from the conceptual inner thrust areas to the outer areas representing the real world of the shop floor and production line. (Courtesy of ICAM, U.S. Air Force, Wright-Patterson AFB, Ohio.)

The robot system is presently being used at General Dynamics Fort Worth Div. fabricating parts for the F-16 fighter aircraft as shown in Figure 8.6. An operator loads 1 of 250 part types onto a rotatable cube work positioner. A minicomputer identifies the part by reading the part number with optical character recognition, and retrieves the corresponding part program from its memory. The minicomputer then orients the

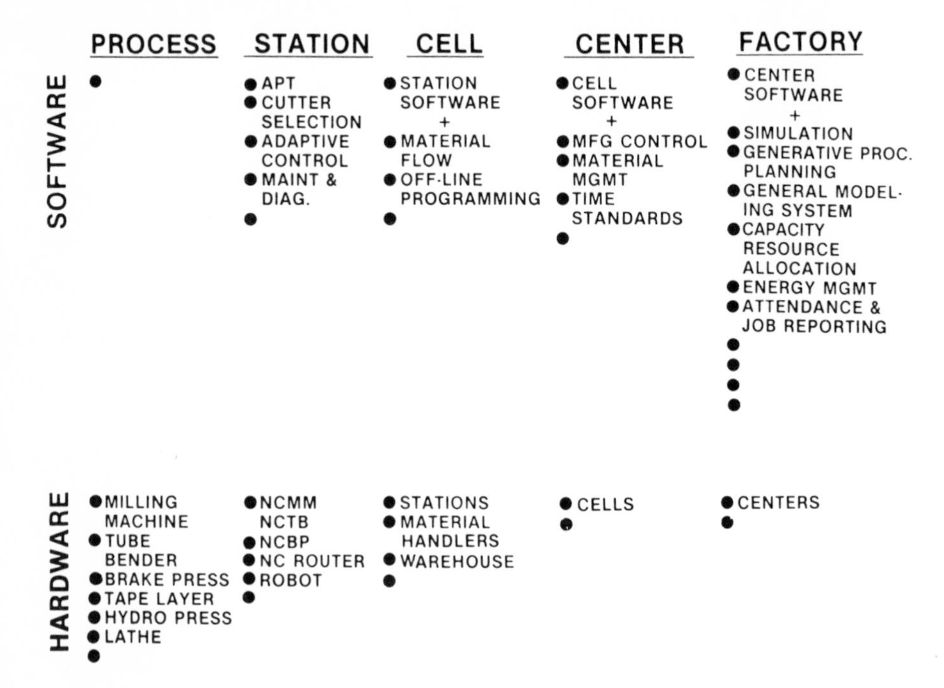

Figure 8.5. The ICAM program concepts are planned to be integrated first at the process and station levels. As these capabilities are demonstrated, real system production processes and stations will be assembled and integrated into a cell function and eventually to the more complex center and factory levels. The lower levels depend primarily on hardware developments, while integration at higher levels depends increasingly on software. (Courtesy of ICAM, U.S. Air Force, Wright-Patterson AFB, Ohio.)

workpiece on the positioner and sends operating instructions to the robot at the operator's go-ahead. If drilling is involved, the robot selects the proper bit from the tool rack and drills holes to 0.005-in. tolerance. The robot then finishes off the part by retrieving a router and completing the periphery of the part. General Dynamics reports production increases of 4 to 1 over manual production methods.

An even more advanced ICAM robotics system is underway at McDonnell-Douglas Co. This effort is aimed at developing a system in which robots, conveyors, and part handlers are controlled with off-line programming, vision through the use of television, and tactile sensors. These capabilities will permit the system hardware to interact with the work-pieces and the control computers.

The ultimate consequence of these ICAM efforts will be the connection of many intelligent robot workcenters into the totally integrated manufacturing system as shown in Figure 8.7. This would produce the real payoff for ICAM and is only now barely within our ability to comprehend, both managerially and technically. In this facility of the future, every phase of design, manufacturing, and management is tied together with computers. This complete integration allows the impact of any changes made any-

Figure 8.6. The 6CM robot arm at work drilling and routing sheet metal access panel for F-16 fighter aircraft. The robot was built by Cincinnati Milicron Co. and purchased by General Dynamics for the ICAM contract. (Courtesy of ICAM, U.S. Air Force, Wright-Patterson AFB, Ohio.)

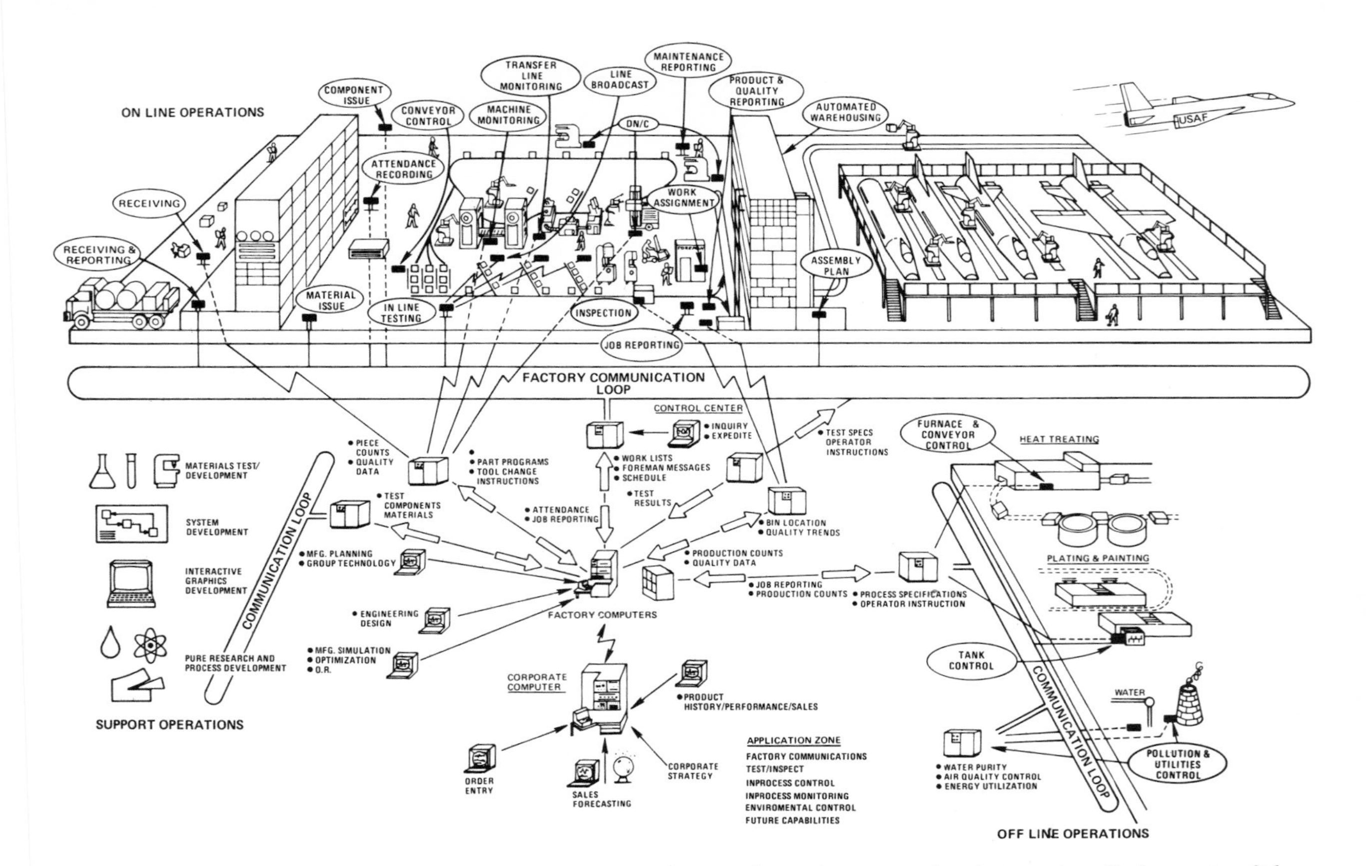

Figure 8.7. ICAM's concept of a computer-managed manufacturing operation integrates all elements of the operation from order entry through automated warehousing and distribution. (Courtesy of ICAM, U.S. Air Force, Wright-Patterson AFB, Ohio.)

where in the factory to be quickly assessed. The result is faster production, efficient performance, and lower cost.

A spin-off of the IPAD and ICAM programs is the Initial Grpahics Exchange Specification (IGES). This will attempt to standardize the communication of geometric data between different CAD/CAM systems as shown in Figure 8.8. Because most present CAD/CAM systems were developed separately in-house or by turnkey vendors, geometric details are organized and transmitted in different formats. As a result, communication between the different systems is plagued with translation problems. The situation has been characterized by one expert as a "Tower of Babel."

IGES will serve as an interim specification until a national standard is developed. Presently, IGES is working with American National Standards Institute (ANSI) and other standards groups. And it is expected that IGES will be the first of many steps required to produce a national standard.

IGES is based on the Boeing CAD/CAM Integrated Information Network, the General Electric Neutral Data Base, and other data-exchange formats. The specification is a set of geometric, drafting, structural, and other entities. Consequently, it has the capability to represent most of the data on a CAD/CAM system.

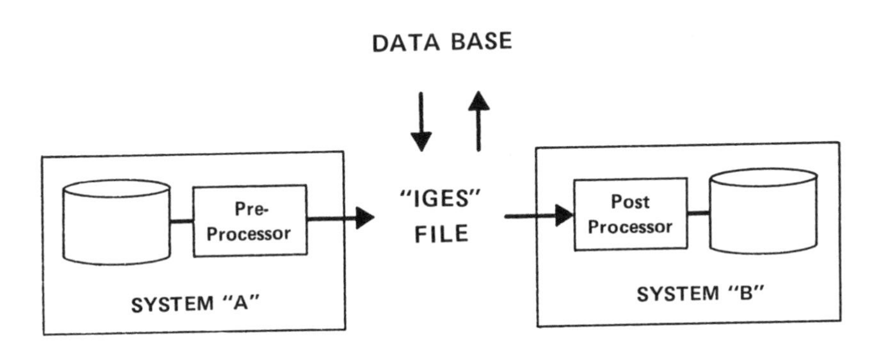

Figure 8.8. This diagram shows the functional description of the IGES concept—an immediate step in the solution of the CAD/CAM data exchange problem. The specification enables geometric data to be transferred between different CAD/CAM systems. (Courtesy of IGES, National Bureau of Standards, Washington, D.C.)

Bibliography

Allan, John J. (Ed.), *A Survey of Commercial Turnkey CAD/CAM Systems,* Productivity International Inc., Dallas, Texas, 1980.

Allen, John J. (Ed.), *A Survey of Industrial Robots,* Leading Edge Publishing Inc., Dallas, Texas, 1981.

Allan, John J. (Ed.), *Management's Guide to Computer Integrated Manufacturing,* Leading Edge Publishing Inc., Dallas, 1981.

Allan, John J. (Ed.), *The CAD/CAM Glossary,* Productivity International Inc., Dallas, Texas, 1979.

CAM-I, "Decision Tools for Manufacturing: From Guesswork to Guidance," Spring Meeting Proceedings, Computer Aided Manufacturing International Inc., Arlington, Texas, April, 1980.

CAM-I, "Man and Computers: Partners in Manufacturing," Winter Meeting Proceedings, Computer Aided Manufacturing International Inc., Arlington, Texas, November, 1980.

CAM-I, "Survival and Growth of the Engineering Industries through Integration of CAD/CAM Technology," Spring Meeting Proceedings, Computer Aided Manufacturing International Inc., Arlington, Texas, April, 1981.

Chasen, S. H., and J. W. Dow, *The Guide for the Evaluation and Implementation of CAD/CAM Systems,* CAD/CAM Decisions Co., Atlanta, Ga., 1980.

Foundyller, Charles M., *Turnkey CAD/CAM Computer Graphics: A*

Survey and Buyer's Guide, Daratech Associates, Cambridge, Mass., 1980.

Fulton, Robert E. (project manager), "Overview of Integrated Programs for Aerospace-Vehicle Design, IPAD," NASA Langley Research Center, Hampton, Va., 1980.

Harrington, Joseph, *Computer Integrated Manufacturing,* Industrial Press Inc., New York, 1974.

IBM, "More About Computers," International Business Machines Corp., Armonk, N.Y., 1974.

IBM, "The Computer Age: The Evolution of IBM Computers," International Business Machines Corp., Armonk, N.Y., 1979.

Krouse, John K., "CAD/CAM: Bridging the Gap from Design to Production," *Machine Design,* June 12, 1980.

Krouse, John K., "Finite Element Update," *Machine Design,* January 12, 1978.

Krouse, John K., "Geometric Models for CAD/CAM," *Machine Design,* July 24, 1980.

Krouse, John K., "Stress Analysis on a Budget," *Machine Design,* March 8, 1979.

Kurlak, Thomas P., "Computer Aided Design and Manufacturing Industry CAD/CAM: Review and Outlook," Merrill Lynch Pierce Fenner and Smith Inc., September 12, 1980, New York.

Machover, Carl, and Robert Blauth, *The CAD/CAM Handbook,* Computervision Corp., Bedford, Mass., 1980.

Nagel, Roger N. (project manager), "Initial Graphics Exchange Specification, IGES," National Bureau of Standards, Washington, D.C., 1980.

Newman, W., and R. Sproull, *Principles of Interactive Computer Graphics,* McGraw-Hill Co., New York, 1979.

Nicholas, J. C., and P. E. Allaire, "Analysis of Step Journal Bearings—Finite Length, Stability," presented at the ASLE/ASME Lubrication Conference, Minneapolis, Minnesota, October 24–26, 1978.

Ryan, Daniel L., *Computer-Aided Graphics and Design,* Marcel Dekker, Inc., New York, 1980.

Wisnosky, Dennis E. (chairman), "The Southfield Report of Computer Integrated Manufacturing," ICAM Project, Air Force Materials Laboratory, Wright-Patterson AFB, Ohio, 1979.

Index